ELEMENTAL

Elemental

HOW FIVE ELEMENTS
CHANGED EARTH'S PAST AND
WILL SHAPE OUR FUTURE

STEPHEN PORDER

PRINCETON UNIVERSITY PRESS

PRINCETON & OXFORD

Published by Princeton University Press
41 William Street, Princeton, New Jersey 08540
99 Banbury Road, Oxford OX2 6JX

press.princeton.edu

All Rights Reserved

Library of Congress Cataloging-in-Publication Data

Names: Porder, Stephen, author.
Title: Elemental : how five elements changed earth's past and will shape
    our future / Stephen Porder.
Description: Princeton : Princeton University Press, [2023] |
    Includes bibliographical references and index.
Identifiers: LCCN 2022042744 (print) | LCCN 2022042745 (ebook) |
    ISBN 9780691177298 (hardback ; alk. paper) | ISBN 9780691248363
    (ebook)
Subjects: LCSH: Evolution (Biology) | Life—Origin. | Ecology. |
    Civilization.
Classification: LCC QH366.2 .P668 2023 (print) | LCC QH366.2 (ebook) |
    DDC 576.8—dc23/eng/20221117
LC record available at https://lccn.loc.gov/2022042744
LC ebook record available at https://lccn.loc.gov/2022042745

British Library Cataloging-in-Publication Data is available

Editorial: Alison Kalett, Hallie Schaeffer
Jacket Design: Katie Osborne
Production: Danielle Amatucci
Publicity: Sara Henning-Stout, Kate Farquhar-Thomson

Jacket Credit: Lee Dalton / Alamy Stock Photo

This book has been composed in Arno Pro

Printed on acid-free paper. ∞

Printed in the United States of America

10 9 8 7 6 5 4 3 2 1

For Beth and Phoebe

CONTENTS

World-Changers

WHAT DOES it take to change the world? Not change in some transient political way, nor even with a revolutionary invention like the wheel, nor the mastery of fire. I mean change in a way that shapes the course of all life on Earth over geologic and evolutionary time. Change like the meteor impact sixty-five million years ago that shrouded the planet in so much dust that it blocked out the sun, wiped out the dinosaurs, and paved the way for the rise of mammals. Such catastrophic, world-changing events don't happen very often. When they do, they shape the tree of life forever.

This book is not about meteor impacts but about different, equally dramatic, world-changing events—events precipitated by life itself. These are rarer than meteor impacts and don't inspire big-budget Hollywood thrillers, but their effects are profound and long-lasting. They arise when evolution produces a new kind of organism, one that can gather certain resources better than any that has come before it. In so doing, such organisms can rework the chemistry of the planet in extraordinary ways.

Such evolutionary leaps are so rare that they are typically separated by hundreds of millions, or even billions, of years. Fascinatingly, a common thread connects these organisms and the changes they precipitate across these unimaginable depths of time. This book explores this connection, tracing the life-driven global change from an anoxic Earth under a faint sun to the industrial world we inhabit today. We'll dive into some of the biggest changes our planet has experienced, the organisms that have caused them, and the lessons we can learn from the past that may help us prepare for the future.

The first and biggest of these changes occurred in a world without animals, plants, or fungi—when no organism was bigger than a single cell. Thus we will start, roughly two and a half billion years ago, with the proliferation of a new kind of single-celled organism called cyanobacteria. These ocean-dwelling microbes created a global environmental catastrophe while setting the stage for the emergence of multicellular life. We'll then jump two billion years forward, to about four hundred million years ago, when the second world-changing kind of organism in our story, land plants, emerged from the water. Their proliferation across the continents took a pan-tropical world and plunged it into an ice age. Lastly, we'll come to the present. As different as humans may seem from cyanobacteria and plants, we have become a third great world-changing organism, and we share much more with our predecessors than meets the eye.

The thread that links these organisms and the changes they precipitate across the unimaginable depths of geologic time is woven from five elements that together make up over 99 percent of every living cell: hydrogen (H), oxygen (O), carbon (C), nitrogen (N), and phosphorus (P). They make up what I'll call "Life's Formula": HOCNP. All organisms great and small engage in a relentless search for these elemental ingredients, gathering them from the environment to build their bodies. Those that succeed—survive. Those that don't—don't. When evolution produces an organism that can gather these elements in a novel, more efficient, and more successful way, it sets the stage for that organism to change the world.

How can the evolution of a single type of organism change the entire world? The answer lies in Life's Formula. In one configuration, these elements are the building blocks of all living matter. In another, they combine to make the gases in the air that keep Earth warm for life to persist (save phosphorus, whose very different role I'll devote a whole chapter to later on). Thus, if evolution produces an organism that can pull unprecedented amounts of one or more of these elements from the environment, the concentration of heat-trapping gases in the air will change, and thus the climate. The more outsized success an organism has in gathering the ingredients of life, the more dramatic climate change will become. In this way these elements link life and climate—in the past, present, and future.

The three world-changing organisms in this story sit on very different branches of the tree of life: microbial, plant, and animal. In part I, we'll dive deep into the geologic past to tell the story of the first two world-changers: cyanobacteria and land plants. We'll explore how the single-celled cyanobacteria evolved new ways to gather the constituents of Life's Formula, particularly carbon and nitrogen, and precipitated the biggest environmental change of all time: the Great Oxidation Event. Then we'll fast-forward almost two billion years and introduce the second world-changing class of organisms—land plants. Their evolutionary innovations in gathering hydrogen, oxygen, and phosphorus allowed them to spread across the previously barren continents. But plants' proliferation inexorably sent the then-tropical planet, with bathtub-temperature oceans at the North Pole, into an ice age that froze many of the world's first forests out of existence. In both cases, we can understand these changes through the elements in Life's Formula, which connect us all to the planet we inhabit. The history of cyano-bacteria and land plants also sets the stage for the story of humans—the third great world-changer.

In part II, we'll focus on how human industry, innovation, and proliferation has precipitated a new geologic era called the An-thropocene. Despite all the obvious differences between us and our plant and bacterial predecessors, we are linked by a common elemental thread: HOCNP. Indeed, this link is the key to unravel-ing the complex web of global environmental woes wrought by modern society. Understanding that is the key to part III, where we'll look to the future. Like our predecessors, our remarkable access to these five elements has brought enormous benefits while pushing us inadvertently toward environmental catastro-phe. Mitigating the unintended consequences of our remarkable innovation depends on our management of the elements we are using to change the world. If we want a more sustainable future, we have a lot to learn from those who came before us.

———

Now that I've briefly introduced our three organisms, let's take a look at the five atoms that make up Life's Formula and shape our climate. Here are two chemical "formulae"* representing two of the world-changing organisms in our story.

$$H_{263}O_{110}C_{106}N_{16}P_1 \qquad \text{Cyanobacteria}$$

$$H_{375}O_{132}C_{88}N_6Ca_1P_1 \qquad \text{Humans}$$

* I put "formulae" in quotes because cells aren't single chemicals; they are mixes of thousands. However, this is the approximate "formula" you would get if you ana-lyzed a whole organism.

For those of you who are a bit rusty with high school chemistry: the letters represent the five elements I've already introduced (plus a sixth, calcium): H = hydrogen, O = oxygen, C = carbon, N = nitrogen, Ca = calcium, and P = phosphorus. The subscripts represent their relative abundance in our respective bodies. For example, in a cyanobacterial cell (the first "formula"), there are slightly more than twice as many hydrogen atoms as oxygen ones (263 to 110), and 263 times as many hydrogen atoms as there are phosphorus atoms. Humans (the second "formula") have a remarkably similar makeup, both in terms of the kinds of elements in our bodies and their relative abundance. Indeed, I could write a "formula" for any living creature, and it would look very similar to these. Of the more than one hundred known elements, these five (plus or minus calcium, for those with bones or shells) are the most abundant in every organism on Earth, in the same order of abundance and with roughly the same ratio. "Life's Formula" is remarkably consistent from bacteria to plants to humans. This shared chemistry puts all organisms in the same boat. All living things need to wring these crucial elements from our environment. To proliferate, life must have access not to one, but to all five—H, O, C, N, and P.

What do organisms do with these elements? Joined in water, H and O make up the vast majority of all cells, and astrobiologists (people who think about and look for life beyond Earth) are convinced that life is impossible without water. To name just a couple of its myriad roles: water is used in photosynthesis, which is the base of almost all food chains (much more on that later), and in the reactions that power all animal cells. Water is easy to get in the oceans, but on dry land staying hydrated is the most pressing need of any living thing. This challenge will play a prominent role when we talk about the evolution of land plants in chapter 2. Carbon, life's third-most abundant element, forms the backbone of all biological molecules: DNA, RNA, protein, fats, carbohydrates, sugars, and many more. Directly or indirectly, most organisms depend on photosynthesis to gather carbon. Photosynthetic organisms (like plants) use the energy of sunlight ("photo") to capture carbon dioxide from the air and synthesize it into biological molecules that store that energy in their chemical bonds. Other organisms (like us and the animals we eat) consume those molecules and break them down to release that energy and fuel our activities, and return carbon dioxide to the air. Thus, photosynthesis directly links life to the most important gas keeping the planet warm—carbon dioxide. I'll explore this link in much more detail in the following chapters. Finally, all living things, photosynthetic or not, need nitrogen and phosphorus to make DNA (and many other key biological molecules). These two elements are embedded in the genetic code of all life on Earth, but as we'll see, they are often in short supply relative to the amount that organisms need to survive.

The abundance of these irreplaceable elements varies enormously across Earth's surface, and this variation dictates where and how much life exists. Little lives in the Atacama Desert in the high Andes of Chile, where it can go centuries without rain. Little lives across vast swaths of Antarctica, where temperatures are too cold for photosynthesis to capture carbon from the air. Perhaps more surprisingly, little lives in the warm, sunlit waters of the central equatorial Pacific Ocean, where a paucity of nitrogen creates a watery "desert." But in the eastern Pacific, where ocean currents enrich the sunlit waters with nitrogen and other nutrients, the sea teems with life. Similarly, on land, in the Sierra Nevada of California, regions underlain by particularly phosphorus-poor rocks host no trees or soil forms, and bare rock abounds. Nearby, where geologic happenstance provides more phosphorus in the rock, giant spruce trees tower above soil-mantled slopes.

The constraints placed on life by the relative abundance or scarcity of these elements are, by themselves, a fascinating story of the world's ecosystems. It is a story explored by my field of science, biogeochemistry, which focuses on how energy and atoms move between organisms and their environments. I've worked for decades in this field, and I'll share some of what I've learned in this book. But as I suggested earlier, these constraints are only half of the story here. These elements are important not just for living things but for the environment in which those things live. They are the constituents of the so-called greenhouse gases that keep our planet warm enough for life to exist. So, let's turn briefly away from the biology of life to the chemistry of Earth.

The main greenhouse gases keeping our planet habitable are carbon dioxide (CO_2), methane (CH_4), nitrous oxide (N_2O), and water vapor (H_2O).* As you can see from their formulae, they consist of four of the five elements that make up all living things. But when these elements are configured in the form of these greenhouse gases rather than living molecules, the building blocks of life create an invisible blanket that traps heat in our atmosphere. As early as the 1850s, scientists had figured out that higher concentrations of greenhouse gas in the air led to a warmer planet. Like glass in a greenhouse or the windows in your car, greenhouse gases allow sunlight to hit Earth's surface, but also trap some of the heat that would otherwise radiate back into space. Glass's transparency to sunlight and capacity for heat retention are why the inside of your car is

---

* I'll try to keep the chemical formulae to a minimum, but I'm going to use CO_2 for carbon dioxide from here onward.

so hot on a sunny summer day. Greenhouse gases' transparency to sunlight and ability to retain heat make the whole planet warmer. In fact, Earth would be a permanently frozen world without them. Greenhouse gases make Earth a Goldilocks Planet—not too hot and not too cold. And the abundance of greenhouse gases in the air is linked, through shared elements, to the activities of living things.

Despite loving science classes in high school, taking lots of science in college (though I was a history major), and even getting a master's degree in geology, I didn't realize that scientists thought about the way biology, geology, and chemistry intersect to shape our living planet until I started my PhD program. Perhaps this is because of the way we teach science, particularly in the United States, where each discipline is taught separately from the others. These silos make it easy to miss the idea that life exists at the intersection of these fields, with biological machinery struggling to overcome the chemical challenge of eking out a living on a rocky planet. Because of these silos, I didn't fully understand what it meant to inhabit a "living planet." Of course, it meant that there is life on Earth. But, just as importantly, it describes a planet *shaped* by life itself, particularly by organisms that can influence the flow of the elements in Life's Formula. I hope that by the end of this book you'll agree that understanding our era of rapid, multifaceted human changes to the planet through this elemental lens offers an illuminating window into those changes and a connection with changes in the past. Perhaps more importantly, I want to show that we can use this way of viewing the world to help navigate toward a more sustainable future.

To set the stage a bit further, I'll return to our three world-changers. Let's start with the cyanobacteria. A little more than two billion years ago, they evolved to combine two very effective ways of gathering and using carbon, nitrogen, hydrogen, and oxygen. First, they used the photosynthetic reaction still used by plants today, which is a very efficient way to use the sun's energy to power biological reactions. Second, they used a process called nitrogen fixation, which allowed them to capture nitrogen from a virtually unlimited supply in the air. As far as we know, no previous organism had combined these two hugely beneficial biochemical processes. This combination gave the cyanobacteria unprecedented access to elements in Life's Formula, which in turn allowed them to increase dramatically in number. We find fossil evidence of their remains billions of years later. However, for those of us who succeeded them on our living planet, their chemical legacy was far more important than their fossils. Their evolutionary innovations increased the total amount of photosynthesis on Earth. And photosynthesis had an unwanted byproduct—oxygen (the molecule O_2—the gas that we need in order to breathe). For two billion years there had been no oxygen in the environment; oxygen was always bound to some other atom (as in water—H_2O). Over time, cyanobacteria pumped out so much oxygen that it overwhelmed the environment's ability to absorb it. This ended the continuous anoxia (lack of oxygen) that marked the first two billion years of Earth's history. Ours became the only known planet in the universe with an oxygen-rich atmosphere, the kind all multicellular organisms (like us) need to breathe. But for the then denizens of the planet, who had been shaped by over a billion years of evolution in anoxic conditions, the transition from no oxygen to oxygen was probably the biggest environmental catastrophe of all time. It fundamentally changed Earth's chemistry, plunged Earth into what was probably its first ice age (more on that in chapter 1), and determined which organisms dominated and which were relegated to the sidelines. All because an evolutionary innovation produced a new way of gathering the elements that shape our living planet and produced a by-product that changed the world.

Two billion years later, an admittedly unimaginable expanse of time, our story's second world-changers evolved: the land plants. They emerged from the water onto the continents around four hundred million years ago and took advantage of a whole new habitat, the 30 percent of Earth's surface that rises above sea level. In order to spread across the land, plants had to evolve ways of gathering three of the five elements in "Life's Formula" that were particularly hard to get in this new habitat: hydrogen and oxygen (in water) and phosphorus. It is no small feat for an immobile plant to stay hydrated (that is, gather H_2O) on dry land. The land plants solved this puzzle by rooting into the underlying rock, creating the world's first soils. These roots pried phosphorus, the fifth element in Life's Formula, from its ultimate source, rocks. Access to unprecedented levels of phosphorus allowed plants to grow as nothing had grown before, creating towering forests on once-unvegetated continents that then spanned from the equator to the South Pole.

The movement of plants from water to land created a host of incidental consequences. The most important aspect of this story is that plants' relentless photosynthesis eventually pulled so much CO_2 out of the air that the blanket keeping Earth warm "thinned" enough to plunge the once-tropical world into a deep freeze. Another near-global ice age ensued. The first tropical forests were frozen by their own success. Once again, evolutionary innovation in gathering the atoms in Life's Formula had catastrophic environmental consequences.

———

At first blush, the changes wrought by humans seem very different from those driven by our world-changing predecessors.

We are sentient, have remarkable technology, and seem so different from cyanobacteria and plants that the thread connecting us with them is not immediately apparent. But if we look a little deeper, as we will here, it turns out that we three world-changers have a lot in common. Human successes and challenges, like theirs, stem from the elements embedded in Life's Formula.

Let me briefly foreshadow human impacts. Every year we burn through hundreds of years of stored photosynthetic energy, the geologically altered cells of our world-changing predecessors, now exploited as fossil fuels: oil, coal, and natural gas. This energy has lifted billions out of poverty, helped increase human lifespans, and (most would argue) improved our quality of life. But releasing this energy has also spewed CO_2 into the atmosphere at a rate that is unprecedented in the last several hundred thousand years, and may well be unprecedented in the history of the world. By the mid-twenty-first century we will likely double the amount of greenhouse gases in the atmosphere relative to the start of the Industrial Revolution in the mid-nineteenth century. And our success is not founded on fossil fuels alone. In less than a century, humans have also doubled the amount of nitrogen in circulation, quadrupled the amount of phosphorus, and captured five times more water in human-made reservoirs than is contained in all the rivers on Earth. These innovations allow us to fertilize and irrigate enough crops to feed our swelling ranks, which as I write is about to exceed eight billion. Even as we reap the fruits of these efforts, the changing stocks and flows of the elements that underlie our success have profound consequences for its longevity. Like our world-changing predecessors, we cannot avoid the elemental links between the living and unliving world. We share the same needs, and as we'll cover in some detail, we share methods to fulfill them with our bacterial and plant predecessors. We should beware of similar consequences.

Despite this rather grim historical perspective, it is critically important to understand that we have two unique advantages when it comes to avoiding catastrophe. Unlike our world-changing predecessors, we can see what is coming. Perhaps even more importantly, we have options for moving away from the way we've done things in the past. We can use this knowledge to inform a transition—one to a society that considers how to manage the elements in Life's Formula in a way that minimizes unintended consequences and maximizes human well-being. We don't know how to do this perfectly, but we know enough to start the transition from managing Earth by neglect to managing it with purpose. For some, the idea of human management of the Earth system is so full of hubris that it's not worth discussing. To this I counter: we are already doing it. Humans are now a dominant geologic force for the global cycles of the elements that change the world. I don't know whether we will act on what we can learn from our predecessors and avoid the worst consequences that come with changing the world. As Yogi Berra is purported to have said, "it's tough to make predictions, especially about the future" [28]. But the only way that we can exit the twenty-first century more sustainably than we entered it is if we act on what we have learned. There is no way back—but there is a way forward.

PART I

Lessons from the Past

1

The Biggest Environmental Change of All

THE HISTORY of Earth is one of environmental change. Dinosaurs have come and gone; so have saber-toothed tigers, mastodons, and myriad creatures that we know only from fossils. The biggest change, however, is not recorded in the remains of animals but in rocks that house no fossils at all. It is a change recorded by the color red.

Perhaps you have seen red rocks in the spectacular cliffs of the Grand Canyon, the towers of Monument Valley, or, as I first did, in Pioneer Valley of western Massachusetts. If you're like me, you paid them little heed. It wasn't until I took a geology class in college that I noticed that rocks even had colors and that most of the rocks around Amherst, Massachusetts, where I was in school, were red.

One of my first geology professors, with the very geologic-sounding name Professor Belt, seemed obsessed with red rocks. He pressed us to think hard about why some were red and others were not. At the time their redness seemed a quirky, unimportant question, and I didn't give it much thought. When we put "Whence the red beds?" on a class T-shirt, it was more a sophomoric jab at Professor Belt's seemingly unwarranted enthusiasm than an attempt to embrace his question. I recall this with regret. I've since come to understand that red rocks are evidence of the biggest environmental change in the history of Earth—and certainly the biggest initiated by its inhabitants. Not surprisingly, Professor Belt knew much more than we did about what was important in the history of our planet.

Sedimentary rocks, the compressed and hardened geologic remains of mud and sand from the ocean floor, are red because they contain iron combined with oxygen (essentially rust). The oldest of these red rocks are about 2.2 billion years old, only roughly half as old as Earth itself. Older sedimentary rocks exist; they just aren't red. In asking "Whence the red beds?," Professor Belt was really asking us "Why aren't older sedimentary rocks red?" The absence of red is a clue, now confirmed by many different lines of evidence, that for the first half of Earth's history, its oceans and air lacked free oxygen. By "free oxygen," I mean oxygen in the form we breathe, which is two oxygen atoms bound to each other and not to anything else (O_2 gas). By contrast, the oxygen in water, bound to two hydrogen atoms in H_2O, is not "free." We can't use it to breathe. In the absence of free oxygen (which I'll just call oxygen from now on), iron doesn't rust, and sedimentary rocks don't turn red.

Today Earth's atmosphere is 21 percent oxygen, but in this, our planet is unique in the universe (as far as we know). Anoxic planets (lacking O_2 gas) are the norm; Venus, Mars, and even planets recently discovered orbiting other stars show no signs of having oxygen. This is because oxygen readily reacts with many things: silicon in the rocks beneath your feet, hydrogen in water, and aluminum in clay. In fact, oxygen bound in this way takes up more volume in Earth's crust than any other element.

Oxygen is also reactive in organisms. When you and I breathe in oxygen, it combines with carbon-based sugars, giving us energy and forming the CO_2 we respire (breathe out). Not dissimilarly (chemically speaking), burning gasoline, a carbon-based molecule similar to sugar, combines that molecule with oxygen in the air and also releases energy and CO_2. Even rotting organic material consumes oxygen: the bacteria and fungi that drive the decomposition of everything from wood to flesh combine the carbon in those substances with oxygen to produce energy.

Given how reactive oxygen is, and with all the breathing, burning, and rotting that happens on Earth, it's worth asking how oxygen ever accumulated in the atmosphere and why it is still around. Shouldn't it all have been consumed long ago? This question arose even before people really knew what oxygen was. In the 1770s several scientists were making breakthroughs that would help answer this question, and the theologian, chemist, and philosopher Joseph Priestley did a groundbreaking experiment. He placed a burning candle in a sealed bell jar and noted that the candle invariably snuffed out. He tried the same experiment with mice and got the same, albeit more macabre, result. He deduced that burning and breathing produced "putrid air." Then he had the real breakthrough. He placed a sprig of mint in the jar of "putrid air" for "eight or nine days" and found that the "sweetened" air was now fine for both the candle and the mouse (Figure 1 [1]).

These experiments, which Priestley published in a series of volumes called *Experiments and Observations on Different Kinds of Air*, were a chemistry tour de force. His meticulous notes are a pleasure to read. But buried in the text, amid myriad detailed observations, instructions, and deductions, is one of the most

FIGURE 1. Joseph Priestley's bell jar, as envisioned by conservation and evolutionary biologist Shahid Naeem. Reproduced with permission.

profound scientific insights I've ever encountered. In ruminating on his results, he notes, almost as an aside,

the injury which is continually done to the atmosphere by the respiration of such a number of animals, and the putrefaction of such masses of both vegetable and animal matter, is, in part at the least, repaired by the vegetable creation. And, notwithstanding the prodigious mass of air that is corrupted daily by the above-mentioned causes; yet, if we consider the immense profusion of vegetables upon the face of Earth, growing in places suited to their nature, and consequently at full liberty to exert all their powers, both inhaling and exhaling, it can hardly be thought, but that it may be sufficient to counterbalance it, and that the remedy is adequate to the evil [1].

I love this passage because it presages our understanding of the Earth system so perfectly. In today's English, and with the benefit of over two centuries of scientific exploration, here's what Priestley was observing: breathing mice, burning candles, and rotting material (he also experimented with rotting mutton and assiduously smelled it over various, and undoubtedly unpleasant, timescales) all consume oxygen and produce CO_2. Photosynthesis, a process that uses the energy from sunlight to capture CO_2 and turn it into biological molecules like sugars, consumes CO_2 and releases oxygen. Though his experiments were performed in glass jars, he realized that Earth is essentially a giant closed container orbiting through space, and that this "putrefaction" and "remedy" must be playing out on a planetary scale. If something wasn't making "sweet" air, it would have long ago been replaced by "putrid" air, as it was in his jar. That would have been very "inconvenient" for animals like us, who need "sweet air" to survive.

Digging deeper into Priestley's passage reveals an even more nuanced insight. He notes that the putrefaction of air can be reversed by the "profusion of vegetables upon the face of the Earth" that are "at full liberty to exert all their powers." Priestley does not explore what might limit their restorative "powers"; he simply states that plants grow where they can best grow. Since Priestley's time we've learned a lot about what limits plants in the exertion of "their powers." Anyone with houseplants knows that they need water, and we've just discussed that they need carbon (in the form of CO_2). It turns out that, in many places, the availability of the other elements (nitrogen, phosphorus) in Life's Formula is what limits plant growth. That's why we fertilize our farms and why, if you've ever seen a sack of fertilizer, you'll see the label "NPK." The N is for nitrogen, and the P is for phosphorus (the K is for potassium).

These limiting elements explain why there is so much variation in the amount of life around the world. Plants need sunlight—that's the "photo" in photosynthesis. But they need more than just sunlight. And the abundance of water (hydrogen and oxygen), nitrogen, and phosphorus varies dramatically across the surface of Earth.

In the rainforests of Costa Rica, where I do much of my research, trees can grow two hundred feet tall and with trunks twenty-five feet across. In the deserts of the midlatitudes, there are vast stretches nearly devoid of plants. To borrow Priestley's phrase, water clearly constrains the ability of plants to "exert all their powers." But even where it is really wet, plants are not free from all constraints. I once took my class and my then nine-year-old daughter Phoebe on a field trip to the central Amazon Rainforest. We spent a week near the confluence of two mighty rivers, the Rio Negro and the Solimões, which come together at Manaus, Brazil, to form the Amazon River. We traipsed through the dark understory, squished through the mud, and squinted up when we heard a bird or a monkey, only to have our views blocked by the seemingly endless layers of leaves stretching far out of sight above our heads. The only views we got were from a tower, built up through the endless layers of leaves to allow measurements (and photos) of the forest below.

One day, however, we were tramping through the forest and found ourselves in the bright sun. The trees were gone, replaced by low shrubs that looked more like they belonged on beach dunes than in the rainforest. I have a great photo of Phoebe sitting on what looks like beach sand, reading her Kindle instead of listening to me wax philosophical about soils and nutrients (Figure 2). She is wearing bright pink sunglasses, boots to protect her from snakes (which don't do much good

FIGURE 2. Sometimes you need shades in the rainforest! Note the bright white (sandy) soils and abundant sunshine—no trees overhead. This spot does not look like your typical rainforest, but without nitrogen and phosphorus, water and sun aren't enough to grow tall trees. Nearby, on richer soils, the trees are a hundred feet tall, and only one to two percent of sunlight hits the ground. Photo by the author.

when you're sitting down), a short-sleeve T-shirt, and *lots* of sunscreen. This may not sound strange, but in all my years of working in tropical rainforests, it's never occurred to me to wear sunscreen—it's very dark under all those leaves. Sunscreen is for days off in town.

We took the class to this spot to contemplate why it was so sunny—that is, why there weren't leaves overhead blocking out the sun. The forest wasn't logged; it was undisturbed. A short stroll away, the trees towered high overhead. Clearly, the difference between the two places wasn't rainfall or sunlight—the tall forest was only steps from the short. The difference was in the soils. In Phoebe's reading spot, the sandy soils, made up of almost pure quartz sand (silicon dioxide, SiO_2), had so little nitrogen and phosphorus that the trees could barely grow on them. Plants on these soils have evolved all sorts of strategies to cope—there are even several species of orchid that are so specialized that they exist exclusively on these sandy soils. Nearby, the more clay-rich soil burgeons with a rainforest that more properly fits our popular imagination of the Amazon—a verdant, towering, and endless sea of green. The fact that soil is important is common knowledge to farmers—some fields are more productive than others because some fields have more nutrient-rich soils, or hold water better, than others. But soils matter for forests, not just for farm fields, because trees, like crops, also are constrained by the elements in Life's Formula.

There are similar variations in the distribution of life across the ocean, where a mélange of single-celled organisms (algae, bacteria, diatoms, and a group called archaea) form the photosynthetic base of the food chain (as plants do on land). The most abundant group of these organisms are cyanobacteria, the world-changers of the deep geologic past and the focus of this chapter. Microscopic, photosynthetic ocean critters are so numerous that together they are responsible for about half of all the photosynthesis on Earth. Yet even in apparently homogeneous water, there are "ocean deserts," where despite abundant sunlight there is very little photosynthesis, and "ocean rainforests" where life proliferates in wild abundance (these also support our most productive fisheries). Satellite observations of the "greenness" of the ocean, which relates directly to how much photosynthesis is happening, show almost as much of a difference in greenness between these regions as there is between the Amazon Rainforest and the Sahara Desert (Figure 3).

FIGURE 3. The "greenness" of the ocean as measured by satellites. The concentration of photosynthetic pigments is represented by white space. The darker the ocean is in this image, the less photosynthesis is happening. Note the vast swaths of ocean "deserts" just north and south of the equator in the Pacific and the Atlantic Oceans. Image courtesy of NASA.

Whence this variation? There is obviously plenty of water. And there is a lot of carbon in the ocean because CO_2 dissolves readily in water, so carbon is not a limiting factor. Just as on land, something else limits photosynthesis. As with our sandy reading spot in the Amazon, it's the other elements in our story, the nitrogen and phosphorus we call nutrients, that make the difference. In ocean deserts, the currents don't supply the nutrients organisms need to flourish.

On land and in the ocean, competition for these nutrients is fierce, and any organism that can get them when its competitors can't has a big leg up in the struggle for survival and reproduction that defines our ever-evolving tree of life. This chapter describes one group of those organisms, the cyanobacteria, and how their evolutionary innovation in gathering carbon and nutrients precipitated the biggest environmental change of all time—the creation of the oxygen-rich planet we inhabit today.

———

Let's return to red rocks and my college geology class. That's where I first learned that, for almost half of Earth's four-and-a-half-billion-year history, the atmosphere was filled with Priestley's "putrid" air. In that world, candles and mice, had they existed, would have immediately been snuffed out because there was no oxygen to "burn," either by combustion or metabolism. It was a very different world, but there were lots of living things. The evidence for the earliest life is indirect, hidden in chemical signatures found in the few rocks from that time still preserved on Earth's surface today. The oldest rocks we've found that contain traces of life are about 3.7 billion years old. While there is still debate about the details, all geologists agree that life has been ubiquitous for most of Earth's history. They also agree that for roughly the first half of Earth's history only single-celled organisms lived in the vast anoxic ocean, with nothing (or almost nothing) on the continents. At first those ocean-dwelling cells got their energy from heat emanating from Earth's interior through thermal vents on the ocean floor. But not long after life emerged, some organisms evolved to use a much more abundant energy source to fuel their growth: the sun. And with the development of photosynthesis, organisms began to change the world.

We don't often describe it as such, but photosynthesis is very much like charging a battery. The process captures energy from sunlight and uses that energy to create chemicals (like sugars) that can be used later to produce energy. Your cell phone bat-

tery does a similar thing when you plug it in—it takes energy
(in this case electrical energy) and uses chemical reactions to
store that energy for later use. While photosynthesis emerged
very early in the history of life, the first photosynthetic organisms used a chemical reaction different from and less efficient
than the one modern plants do. The earliest photosynthesizers
were like the Model T car. Their evolution was revolutionary
but with lots of room for improvement. There are still organisms that use that "primitive" photosynthetic process: the so-
called purple sulfur bacteria (they can also be pink, orange, or
black). Today they make their living in anoxic but sunny environments such as salt marsh mudflats. On early Earth they were
everywhere. But now, as then, they are not particularly efficient
at converting sunlight into stored energy. Put another way, for
a given amount of sunlight, they cannot make very much sugar,
so their chemical batteries aren't so great.

Sometime after the earliest forms of photosynthesis evolved,
perhaps as early as 3.5 billion years ago, perhaps as late as 2.7
billion years ago, a new "modern" photosynthetic process
evolved ("modern" because this type of photosynthesis is dominant on Earth today). If the photosynthesis of the purple sulfur
bacteria was the Model T of CO_2 capture and energy storage,
modern photosynthesis is closer to a Tesla. In the primitive process, energy from the sun is used to split molecules of hydrogen
sulfide (H_2S) to build their chemical batteries. Modern photosynthesis uses water molecules (H_2O) rather than hydrogen
sulfide. Not only is water more abundant than hydrogen sulfide
(water is, of course, the most abundant chemical in the ocean),
but for a given amount of sunlight, modern photosynthesis can
create more sugar—more stored energy. I've written out the
chemical reactions for primitive and modern photosynthesis
here. You can see that they are similar—they both use the energy

of sunlight to capture carbon from the air (in the form of CO_2) and store that carbon in a biological molecule (sugar) for later use. They also both produce waste products—sulfur in the first case, oxygen in the second. We'll come back to that difference later in this chapter.

Primitive photosynthesis

carbon dioxide + H_2S + sunlight → sugar + S + water

Modern photosynthesis

carbon dioxide + H_2O + sunlight → sugar + O_2

The evolution of modern photosynthesis gave the organisms that could use it an incomparable advantage. The advantage was so big that this innovation likely increased the total mass of life on Earth by a factor of ten, though, as with anything that happened so long ago, there are a lot of debate about the details. In the context of this story, there is a question that we need to consider. If this new way of doing photosynthesis was so great, what, if anything, limited these innovators? After all, they had invented a more efficient way of capturing carbon and storing it in their chemical batteries. Given the ability of photosynthetic organisms to use sunlight (of which there is plenty) and water (of which there is plenty) to make biological molecules, why didn't the ocean turn into a giant, salty pond filled with green, photosynthetic "Teslas"? This question has profound implications for life on the early Earth and life today.

To stretch the metaphor: those evolutionary innovators were probably limited by their battery factories. They had a great product, but they needed the raw materials to scale up production. Having an idea for better batteries doesn't do a lot of good if you don't have the materials to build them. So what specifically, besides sunlight, water, and CO_2, might have held these photosynthetic Teslas back?

The answer lies in the elements of Life's Formula. The ocean is a great source of H and O, and sunlight plus photosynthesis provides carbon. But what about the cellular machinery that carries out photosynthesis? It turns out this machinery uses a lot of nitrogen. And it's likely that in the early oceans, nitrogen was the raw material that limited the production of those photosynthetic Teslas. For every six atoms of carbon, photosynthetic organisms in the oceans contain one atom of nitrogen. Having spent some time on the evolutionary leap that produced a new way to capture carbon, I want to turn now to a second innovation of the cyanobacteria, a new way of capturing nitrogen. This requires a dive into the chemistry of nitrogen, perhaps the most mercurial of life's regulators. Once we understand the basics, we can understand the biggest environmental change of all time.

———

I once saw a lecture by one of the modern leaders in the study of nitrogen and how humans use it, James Galloway. I always like science talks that link to the world of the arts and thus was intrigued when he introduced the idea of nitrogen scarcity with a reference to Samuel Coleridge's "The Rime of the Ancient Mariner." The idea of "water, water, every where / nor any drop to drink" [29] aptly describes sailors dying of thirst in a sea of salt water (Figure 4). It is also an apt metaphor for the plight of all life on Earth with respect to nitrogen. Nitrogen gas makes

FIGURE 4. Nitrogen, nitrogen, everywhere, nor any an atom to use. Nitrogen in the air, like salt water, needs to be transformed before it can be used. Gustave Doré's 1876 illustration of Samuel Coleridge's "The Rime of the Ancient Mariner." Image courtesy, iStock; Duncan, 1980.

up almost 80 percent of the air. We are constantly surrounded by it, breathing it in and out. But that breathing cannot capture nitrogen in a form our bodies can use, despite its appearance in Life's Formula. In fact, very few organisms can capture nitrogen and use it. Most just take it in and release it unchanged—and useless.

The reason nitrogen is so hard to get is that nitrogen in the air is an almost-inert gas: two nitrogen atoms bound by a very strong triple bond (N_2). Unlike the two oxygen atoms in O_2, which readily split apart and react with other elements, the atomic configuration of nitrogen makes the N_2 molecule a very difficult nut to crack. Unless the strong bond between the two nitrogen atoms in N_2 can be broken, N_2 is "biologically unavailable." That's jargon for "useless." The stuff we can use, "available nitrogen," is nitrogen in which that strong bond has been broken and one nitrogen atom is attached to something else (usually hydrogen, oxygen, or carbon). The vast majority of nitrogen in the world is biologically unavailable to us and almost all other organisms. *Almost* all organisms. I'll come back to that shortly. First, let's go back to early Earth and the cyanobacteria.

Some available nitrogen existed in the early oceans as a remnant of the formation of the planet, but it was probably very scarce. Lightning heats the air so much that it can break the bonds of an N_2 molecule and allow a single nitrogen atom to combine with something else and become available. Lightning was probably the main external supply of available nitrogen to the early ocean. We don't know a lot about how much lightning there was three and a half billion years ago, but there almost certainly wasn't enough to make a lot of available nitrogen. To make matters worse for organisms trying to wring nitrogen from their surroundings, what scarce nitrogen was available was probably shuttled quickly back to the atmosphere, because there were (and still are) microorganisms that got their energy from changing available nitrogen back into N_2 gas. Thus, the amount of life in the early ocean was profoundly limited by the amount of available nitrogen. To borrow a phrase from Joseph Priestley, even organisms that could use modern photosynthesis couldn't "exert all their powers."

If the development of photosynthesis was the most important evolutionary step ever taken by life on Earth, a close second was the evolution of an enzyme that allowed a few select organisms to break apart N_2 gas and turn it into available nitrogen. This process is called biological nitrogen fixation, and the enzyme that does it is called nitrogenase ("nitrogen breaker"). As with modern photosynthesis, we have only a rough estimate of when the first organisms evolved to produce nitrogenase, and this period spans the same billion-year timeline of three and a half to two and a half billion years ago. Sometime around that later date, however, these two world-changing reactions, photosynthesis and nitrogen fixation, were brought together in a single group of organisms—cyanobacteria. The first world-changers had emerged.

Cyanobacteria had the complete package. They had a process to build a more efficient battery—capturing and storing sunlight energy in carbon-based sugars. And they had another process that allowed them to build more factories to churn out those batteries, because they had a way to take inert nitrogen from the air and capture it for their own use. Access to carbon and nitrogen together allowed cyanobacteria to proliferate wildly. They proceeded to take Earth over a precipice of biologically driven change unseen before or since.

Our best guess is that cyanobacterial innovation prompted another tenfold increase in the mass of life on Earth, over and above the increase that followed the evolution of modern photosynthesis by itself. Equally important to our story is that the success of cyanobacteria meant modern photosynthesis became an even more ubiquitous chemical reaction. Just so you don't have to flip back a few pages, I'll repeat the reaction for modern photosynthesis here:

$$\text{carbon dioxide} + \mathbf{H_2O} + \text{sunlight} \rightarrow \text{sugar} + \mathbf{O_2}$$

Modern photosynthesis breaks water apart to capture CO_2 and make sugars, releasing the waste byproduct, oxygen, in the process. Cyanobacteria have no use for this byproduct. In fact, oxygen inhibits both photosynthesis and nitrogen fixation, interfering with the enzymes that makes those reactions happen, though not enough to outweigh the benefits to the cyanobacteria. Cyanobacteria may be the first example in Earth's history—but, as we'll see, not the last—of a wildly successful organism whose waste product would change the world.

At first, the oxygen cyanobacteria produced reacted with other abundant elements in the ocean. Eventually, however, the combination of high oxygen production, the reaction of that oxygen with the vast, but not infinite, supply of things like iron and sulfur in the ocean, and geologic happenstance allowed oxygen to saturate the ocean and percolate out into the air. A little less than two and a half billion years ago, oxygen levels in the air reached unprecedented (meaning detectable) levels. I'll briefly describe this extraordinary Great Oxidation Event here. If you're interested in more details, Donald Canfield's *Oxygen* has much more on this topic [2].

In today's world, all the life we can see depends on having oxygen in the air. But two billion years ago, its accumulation in the ocean and air was an environmental catastrophe. In order to understand why, we need to walk through what Earth was like at the time in a little more detail.

On early Earth, the young sun was only about three-quarters as bright as it is today, which meant that there was less incoming energy to warm Earth's surface. If it hadn't been for a strong greenhouse effect, the sun would have been too dim to unfreeze the oceans. I talked about the greenhouse effect in the introduction, and I'll briefly recap it here. Some gases in the air, the greenhouse gases, are transparent to sunlight, which passes through those gases and warms Earth's surface. But these greenhouse gases absorb some of the infrared radiation that the warmed surface emits. The greenhouse gases thus act as an invisible blanket, trapping heat and keeping Earth warmer than it would be in their absence. There are four main greenhouse gases: CO_2, methane (CH_4), nitrous oxide (N_2O), and water vapor (H_2O). Today, CO_2 and water vapor are responsible for the majority of Earth's greenhouse effects.

On early Earth there probably wasn't as much CO_2 in the air as there is now (CO_2 is stable when oxygen is high, but not when it's not). Nevertheless, the greenhouse effect was strong because there were relatively high levels of methane in the air, and methane is a very effective greenhouse gas—actually more effective than CO_2 at trapping heat. Methane formed a blanket warm enough to keep the oceans unfrozen for almost two billion years. But methane, unlike CO_2, is not stable in the presence of oxygen.

Into this world warmed by methane came an evolutionary accident. Cyanobacteria combined a very efficient form of photosynthesis that happened to produce oxygen with a capacity to capture nitrogen that allowed them to do a lot of photosynthesis. They became the dominant form of life in the ocean, forming huge coastal colonies which we find as fossils today. Over tens, or maybe even hundreds, of millions of years, they went about their business, capturing (or fixing) nitrogen, using that nitrogen to build the photosynthetic machinery that turned sunlight into chemical energy, and turning that energy into more cells of cyanobacteria. Their way of living released oxygen, an unwanted byproduct.

At first, this didn't really matter. For a very long time, Earth seems to have been unaffected by this new pollutant, which reacted quickly with all sorts of other atoms in the ocean. When oxygen reacted with iron, for example, it formed the red rocks that so fascinated Professor Belt. However, as cyanobacteria continued to pump out oxygen, eventually enough accumulated that it began to bubble out of the ocean and into the air, where it destroyed most of the methane that was keeping the planet warm. When that happened, about 2.2 billion years ago, the planet plunged into an ice age that may have lasted millions of years.

This Great Oxidation Event was thus a slow-moving double whammy. First, many of the denizens of the ocean were obligate anaerobes, meaning they couldn't carry out their basic life functions in the presence of oxygen. After almost two billion years of living in an anoxic world, they were faced with a toxic pollutant (oxygen) and thus began their relegation to a much more limited habitat. Second, the Great Oxidation Event weakened the global greenhouse, perhaps sufficiently to induce a global, or near global, glaciation, creating the first Snowball Earth (a planet completely coated in ice), though the timing and link between this oxygenation and glaciation are still debated. As ice spread from pole to pole and anaerobic organisms lost most of their habitat, the Great Oxidation Event caused a giant deflection in the trajectory of life on Earth. It was also the first world-changing event and environmental catastrophe precipitated by life itself.

We have no idea how many organisms went extinct as a result of the Great Oxidation Event. The geologic record from that time is sparse. All life on Earth was single-celled and thus not likely to be preserved in the fossil record. What we do know is that organisms that made a living in a world with no oxygen once dominated the oceans and that many of their descendants are now relegated to patchy geologic oddities like sulfur-rich hot springs or volcanic vents on the sea floor. Conversely, the accumulation of oxygen set the stage for a world in which multicellular life (including humans) could emerge. Life can thrive in all sorts of conditions. But living through the transition from one state to another is neither pleasant nor guaranteed.

The Great Oxidation Event precipitated by cyanobacteria has another important lesson for us. If an organism evolves a way to alter the flows of even a few of the five elements in Life's Formula, they can proliferate. Because those elements are also tightly linked to climate, a change in this flow can change the world. Indeed, the flow of these elements between the living and unliving world, their so-called biogeochemical cycles, are the true chemical pulse of the planet. As with cyanobacteria, when an organism proliferates to the point where it changes the amount of greenhouse gases in the air, the changes can be dramatic. Typically, what stops such runaway proliferation is limited access to nutrients (we've already discussed nitrogen) and, if those organisms are on land, limited access to water. If limits on those other regulators can be overcome, as with cyanobacteria, really big changes commence.

The second and third parts of this book are about humans, their emergence as a world-changing organism, and the ways in which we can build a more sustainable future. But I chose to start with cyanobacteria, which preceded us as world-changers by billions of years, because their story is so intimately tied to ours. There is the obvious—no multicellular life, including humans, could exist without oxygen in the atmosphere. We have cyanobacteria to thank for making the planet habitable for us. Equally important, if more subtle, is the link between Life's Formula, its five elements, and environmental change. Remember that efficient photosynthesis, which used sunlight to build carbon-based chemical batteries, evolved long before oxygen accumulated in the atmosphere. Nitrogen fixation likely evolved before that. But when organisms combined these two innovations—better carbon capture and the ability to fix their own nitrogen—their waste product (free oxygen) changed the world. Moreover, the collateral damage of their success was a global glaciation that likely lasted millions of years. Imagine what a challenge it must have been to survive—going through the transition from a warm, watery, anoxic world to an oxygenated, periodically frozen Snowball Earth. It took a long time and some geologic happenstance for oxygen-producing photosynthesis to change the world. But once it did, Earth never went back. Life changed the planet forever.

Earth is a "living planet," but it is rare for a single group of organisms to profoundly change the world. Before we get to the third world-changing organism (humans), I want to explore how and why the second, land plants, were able to do something similar to their cyanobacterial ancestors. Like cyanobacteria, land plants evolved a new way of getting carbon from sunlight, this time by taking the same chemical reaction to a new place (land). On land they became the world's first, and most efficient, miners. Their roots excavated unprecedented quantities of other key elements (namely phosphorus) that allowed them to proliferate wildly. For over a hundred million years, these innovations allowed land plants to enjoy unparalleled success. Then, inexorably, their successes precipitated another global environmental catastrophe.

2

Plants Colonize the Continents

AS A GRADUATE STUDENT, I had the enviable task of doing my dissertation work in Hawai'i. On one of my first days there, my PhD advisor Peter Vitousek and his wife Pamela (Pam) Matson took me on a hike to look at an active lava flow as it poured into the ocean. You could see towering jets of steam as lava met water far to our right—but we were angling uphill, away from the toxic chlorine gas and shards of volcanic glass spewed out by the collision of these geologic forces. Nevertheless, his warnings that we would have to "get the hell out of there if the wind changed" were enough to keep me on my toes. Stumbling over freshly cooled lava, which is jagged enough to shred your boots pretty quickly, was also a bit intimidating. And keeping up with Peter, an oft-discussed rite of passage in the lab, was no easy task. Rumor had it he could chase and catch feral pigs across a lava landscape while wearing only flip-flops on the sharp *a'a* (lava so named for the sound you make when walking across it).

As I paused at one point to catch my breath, Peter tapped me on the shoulder and said, "Look down." Beneath my feet, glowing up through a crack in the just-cooled surface, was bright orange lava, radiating heat I hadn't noticed in my attempt to keep up. I was too nervous to think about science much at the time, but as I think back to that day, I realize it was the only time I've been on lifeless land.

Freshly cooled lava is one of the only surfaces on land unaffected by living things. Even the soil in the driest desert in the world teems with bacteria and other small critters, as do the coldest, driest valleys of Antarctica. In lush Hawai'i, it takes only a year or two before lava flows are covered in white fuzz (lichen that form symbioses with nitrogen-fixing bacteria), and in a few more years, the lichen give way to plants. Trees pop up from cracks in the ground after a decade or two, and in a century, full-fledged forests have emerged atop a thin layer of soil.

Plants are so ubiquitous on land that it's hard to imagine their absence, but for almost 90 percent of Earth's history, there was no life on land or at the very least no plants. Land plants emerged a little more than four hundred million years ago, fully two billion years after cyanobacteria oxygenated the planet. As far as we know, no other world-changing group of organisms has left as much of a mark in that two-billion-year interval, though, as always with the deep geologic past, there is some debate.

By the time plants moved onto land, the oceans from which they emerged were not radically different from those that cover 70 percent of Earth's surface today. Earth was much warmer, with a tropical climate from pole to pole, but all the major types of life, including vertebrate and invertebrate animals of all shapes and sizes, roamed the seas. The continents were about their current size, though not in their current location. Ocean food chains were rooted in oxygen-producing photosynthesis, sustained by nitrogen fixation.

Before we take a closer look at the jump to land and how it fits into the story of world-changing organisms, I want to set the stage by exploring the oceans just a little bit more. Specifically, I want to talk about the supply of the elements in Life's Formula, because doing so sets the stage for the remarkable revolution marked by the emergence of plants onto the continents.

Four hundred million years ago, as today, and probably for much of the history of the ocean, nitrogen placed a key limitation on how much life existed in the ocean. I didn't focus on this in the last chapter, but it is a bit of a puzzle as to why this would be the case. Cyanobacteria had evolved the ability to fix nitrogen, importing it from the virtually limitless pool in the atmosphere (nitrogen gas readily diffuses into ocean water). This gave cyanobacteria a competitive advantage—they could get a key nutrient that others couldn't. But once a cyanobacteria cell dies and decomposes, the nitrogen it captured should become available to other organisms. Recycling is the norm in nature—once a scarce nutrient enters a system, it tends to stay there, fiercely sought after by all concerned. So why, when cyanobacteria could tap into a virtually limitless bank account of nitrogen in the air, did nitrogen remain relatively scarce in the ocean? Why didn't the cyanobacteria cause it to accumulate until it was no longer in short supply?

This puzzle has preoccupied scientists in my field for decades, and like many good puzzles there is no single, clear answer [3]. Here, I want to focus on one among many reasons people have come up with: that the proliferation of cyanobacteria specifically, and photosynthetic organisms in the ocean in general, are limited by another element in our story.

The idea that an organism might evolve a better way of getting one of life's essential elements only to be limited by another should ring a bell. It's at the heart of the last chapter. The evolution of a better way to capture carbon (modern versus primitive photosynthesis) was a leap, but the leap was augmented when modern photosynthesis was coupled with nitrogen fixation. In this chapter, let's take that idea a step further. Cyanobacteria have excellent machinery for gathering carbon and nitrogen. What limits their growth?

Let's go back to the elements that connect all world-changers: H, O, C, N, and P. Living in the ocean, cyanobacteria had plenty of access to the first two, and indeed modern photosynthesis uses them in a very efficient way. There is no shortage of carbon in the ocean, and research from as early as the 1950s showed convincingly that enough CO_2 gas dissolves into the ocean that it rarely, if ever, is a constraint to growth [4]. Cyanobacteria can fix nitrogen, which also dissolves in ocean water since it is so abundant in the air. I ended the last chapter by saying cyanobacterial innovations in carbon and nitrogen capture changed the world, and that's true. But I confess that was also a bit of a sleight of hand because I left out the last, critically important element: phosphorus (P).

It turns out that organisms that can fix nitrogen tend to have high demands for other atoms—particularly phosphorus, but also iron and molybdenum. The latter two are important components of the biological machine (the nitrogenase enzyme) that carries out nitrogen fixation. Phosphorus, iron, and molybdenum, unlike nitrogen, are virtually absent from the air. They are made available to organisms by the chemical breakdown of rock, and thus, with an admitted lack of linguistic imagination, scientists call them "rock-derived." We now think that these rock-derived elements might limit the growth of cyanobacteria and other nitrogen-fixing organisms in the oceans. Thus, while life might have been proximately limited by the amount of nitrogen, the amount of nitrogen those organisms could capture was ultimately limited by the supply of elements derived from the weathering of rocks.

Imagine yourself as a single-celled, photosynthetic organism floating in the middle of the ocean, more than a thousand miles from land. If you're at the surface, there is plenty of sunlight available to drive photosynthesis. There are plenty of water molecules to split using the energy from the sun. If you're a nitrogen fixer, you can build the machinery to capture nitrogen gas that is dissolved in the water, so you can build the enzymes to carry out those processes. But where do you get the elements—the rock-derived phosphorus, iron, and others—needed to build that machinery? Not from the weathering of rocks at the ocean bottom—those are miles down—and even if you managed to get down there, there wouldn't be any light to fuel photosynthesis. As a single-celled organism in the upper ocean, you would just have to wait and hope that those elements come to you.

If you're lucky, you're in a place where currents bring nutrient-rich ocean water to the surface. Life abounds in places like these. It's where the majority of fish live, because it's where the majority of photosynthetic organisms (fish food) live. But if you're an unlucky single-celled organism, you live in a vast ocean desert, like the central Pacific I described earlier in the book. These places have very little life, despite being replete with sunlight and CO_2, because they lack the other parts of Life's Formula. The only source of rock-derived elements is the transport of material from the continents—a slow trickle of dirt from rivers and dust falling on the ocean surface. You are at the mercy of the currents. There are no rocks for miles: up, down, or sideways. There is nothing you can do to increase your access to rock-derived elements. No way to access the fifth-most abundant element in your cells—phosphorus—and the other atoms derived from the breakdown of rocks. No way, that is, except to evolve and move to the source: land.

———

For more than three billion years, life was mostly restricted to the oceans. While evidence is emerging that there was some life on land for much of that time, it was apparently restricted to mats of desiccation-tolerant organisms that rose only millimeters above the surface. The effect of these "cryptic crusts" on the Earth system is hotly debated, and as is typical of organisms with the word "cryptic" in their description, they remain mysterious to us. But scientific consensus has emerged about the arrival of plants on land—consensus that a second great change initiated by evolutionary innovation had begun.

As with the cyanobacteria before them, the innovations that allowed plants to complete the slow march landward revolved around access to the elements in Life's Formula. A first, and critically important, step was to bring the photosynthetic machinery from the ocean with them. The chloroplasts in plant leaves—the place where photosynthesis occurs—have their own DNA. It's the DNA of photosynthetic ocean bacteria that, long ago, merged into plant cells. Chloroplasts are thus an example of endosymbiosis—an organism within an organism. As a result of this endosymbiosis, the chemical reaction of plant photosynthesis is the same as cyanobacteria photosynthesis. It uses the same machinery. That is why land plants pump out oxygen during photosynthesis in the same way cyanobacteria do.

The biology of endosymbiosis is amazing and has produced some of the most important evolutionary changes of all time. Mitochondria, the "power plants" that fuel human and plant cells, are another example of an endosymbiosis. Modifications through endosymbiosis were long overlooked by evolutionary biologists until the brilliant Lynn Margulis provocatively (and correctly) claimed that they were responsible for some of the most important evolutionary events in the history of life. She was decades ahead of her time, but modern genetics has proven her right.

Land plants brought photosynthetic machinery with them, but life on land was still a tricky business for organisms made mostly of water. Rain is sporadic, even in places that get plenty of rain. Swampy lowlands are the easiest places to get a foothold, and the fossil record suggests that that is where plants remained for tens of millions of years. This could be a bias in the fossil record—swamps are excellent for preserving fossils because sediment accumulates quickly, and then low oxygen in the thick mud inhibits decomposition and favors fossil preservation. But there are other reasons to suspect that the transition out of the wet lowlands might have been slow and evolutionarily difficult. Before the emergence of land plants, soils were likely thin, if they existed at all, and not particularly good at holding water. Fossils of the first plants show only shallow, superficial roots, and thus indicate that they were not likely to be successful in dry, rocky soils.

The first abundant plants to emerge were likely the ancestors of liverworts, primitive plants that still exist today, but the surviving groups are small and somewhat nondescript. Think "leafy-looking moss," though my botanist friends would cringe at mixing these two terms. Structurally, though, the first big plants were quite different from today's liverworts. These early colonists grew almost as tall as today's trees but were more like photosynthetic telephone poles. They were "trees" with no branches except at the top, and no top until the plant produced reproductive tissue and then died. Seeds, which as we will see were critical for the transition to drier areas, did not evolve for another fifty million years.

Living in the ocean meant using water for photosynthesis wasn't a problem. But on land, the need for water means a constant struggle for plants to stay hydrated. The struggle is encapsulated by Life's Formula, which starts with H and O. Because land plants inherited their photosynthetic machinery from their ocean-dwelling, single-celled ancestors, they use the hyper-efficient, water-reliant modern photosynthesis we've already discussed. They split water using the energy from sunlight, capture CO_2, and produce sugars to build their cells (and oxygen, by evolutionary accident). What's more, as I'll describe in a later chapter, every moment they let CO_2 diffuse in from the air is a moment when they are losing scarce water out through the same conduit. This is a scarcity ocean-dwellers don't have to deal with.

Despite posing the challenge of staying hydrated, life on land allowed plants unprecedented access to the other elements in our story. Plants had abundant sunlight—an energy source that fueled their access to CO_2 in the air. But building photosynthetic machinery needed a lot of nutrients: not only nitrogen but rock-derived nutrients such as phosphorus and iron. On land, there was no need to wait for rivers, dust, or ocean currents to bring those elements to them. Land plants moved to the source. They began to crack apart the ground for unprecedented access to the elemental goodies beneath their roots.

The timing of the evolutionary innovations that allowed the transition to land is only partially understood. Probable fossil remains of flattened plants have been discovered in rocks a bit more than four hundred million years old (in Virginia, Libya,

and Wales, among other places), close in time to the first appear-
ances of fossil roots and the first putative fossil shrubs. By three
hundred seventy-five million years ago, however, fossils of large,
tree-like plants were widespread. Called *Archaeopteris,* they grew
up to one hundred feet tall, had trunks five or so feet wide, and
had leaves resembling the fronds of modern ferns. *Archaeopteris* wood was almost indistinguishable from that of modern
conifers (pines, spruces, and the like). Fossil roots reveal that
belowground they also looked a lot like modern plants. They
formed deep, branching root systems, and those roots formed
associations with fungi (80 percent of all modern plants also
have fungal partners). As fungi do today, those ancient fungi
almost certainly provided water and nutrients to plants in exchange for carbon derived from photosynthesis. Thus, while
Archaeopteris forests were in many ways quite different from
today's forests, and while some scholars rebuff the idea of calling
these early ecosystems "forests" at all, their similarities are more
important for our story than are their differences.

By colonizing the continents and moving to the source of
the elements whose availability constrained their ocean-
dwelling ancestors, land plants set themselves up to become
the second great world-changers. Let's take a quick tour of the
world they eventually changed. The atmosphere plants encountered when they emerged on land was in many ways similar to the air today—nitrogen (as N_2 gas, two nitrogen atoms
bound so tightly together they are virtually inert) and oxygen
(as O_2 gas, two oxygen atoms bound together loosely enough
to be very reactive) made up the vast majority of the air. But
the best available evidence suggests CO_2 levels may have been
ten times higher than today, and the heat trapped by all that
CO_2 meant the world was very hot, probably about 10°F hotter
than today. This may not sound like a lot, but such a world was

hot enough to have no ice at either pole. The surface of the ocean may have been as hot as 100°F (that's hot tub temperature), and it would have been possible to swim without a wet suit at the North Pole. Ocean waters covered almost the entire northern hemisphere, while the supercontinent Gondwana draped over the South Pole.

I want to reiterate the challenges faced by an immobile organism (like a plant) on land. Like any living thing, plants require a suitable habitat, and there is no guarantee that wherever they end up will be suitable. Thus, movement is key. Ocean dwellers depend on currents for transport. While temperatures, salinity, and available nutrients may differ among oceans, wherever these currents take ocean dwellers, they will stay in water. Access to at least two of the most abundant elements in Life's Formula, H and O, is guaranteed. On land there is no such promise. Land is highly heterogeneous. Some hills face south, so in the northern hemisphere they tend to be hotter and drier than hills that face north. Some soils are easy to root in, others are not, and soils differ wildly in how well they hold water and nutrients depending on what rocks they are derived from. Colonization of land requires, first and foremost, an ability to withstand water scarcity, something never encountered in three billion years of life in the ocean.

Several key evolutionary innovations related to water emerged over the tens of millions of years between the emergence of the first swamp-dwelling land plants and the emergence of the first forests. Leaf waxes helped limit the loss of water, elongated water-transporting cells facilitated the movement of water through the plant, and rigid structural molecules called lignin and cellulose allowed *Archaeopteris* to stand tall. There is some debate about how important competition for light was in these early forests, but the rigidity provided by lignin also prevented the vessels that transported water from the roots to the leaves from being destroyed when a tree swayed in the wind. Maybe even more importantly, rigid materials also allowed roots to push their way through the rock and developing soils in the search for water and nutrients.

The power of plant roots is really amazing. Outside my house a tree root is busy breaking apart my sidewalk. One of my PhD research sites in Hawai'i sits on a three-hundred-year-old lava flow, and roots already have helped create almost two feet of soil atop that recently deposited lava. Near this site is a lava tube—a cave left behind after the lava flowed out—and although it's twenty feet or so beneath the surface, there are roots growing down through the ceiling of solid rock. Roots have so much power that they have a profound effect on their environment. They physically break apart rocks, including the cement in my sidewalk, by prying them apart. They also chemically attack them by releasing acid. This interaction between roots and rock turns out to be a critical component of how plants changed the world. As we will see in the next pages, this importance is only partly because roots provided plants with access to water. There is more to the story.

Remember that water is necessary, but not sufficient, for an organism to proliferate and change the world. Plants had another puzzle to solve—nutrients. And roots helped overcome that constraint, too. If limited water is the biggest drawback of living on land, by far the biggest upside is access to nutrients that are stored in rocks but scarce in the oceans. Rocks and the nutrients they contain are right beneath the surface. By developing roots and connecting with fungi that form miles of microscope filaments that permeate every nook and cranny, land plants became the first miners in history. Just like human miners, land plants start by physically prying the rock apart, and then they use chemistry to turn the rock into something useful. Scanning electron microscope images of minerals in the soil reveal tunnels and etchings that are "scars" from the relentless attack of plants and their symbiotic fungi in the search for nutrients. Even the rotting of dead roots and fungi releases CO_2 (Priestley's "putrid air"), and this CO_2 acidifies the water in soil, helping to dissolve rocks further. With roots and their associated fungi, land plants evolved a way to mine rocks for nutrients. And the concentration of phosphorus, our focus here, is hundreds or even thousands of times higher in rocks than it is in the ocean.

By a little less than four hundred million years ago, access to water and nutrients had allowed the first *Archaeopteris* forests to spread from the equator to the South Pole (there was little land north of the equator). But it is difficult to tell how much land they occupied. Their fossils are mostly restricted to what we interpret to be low-lying wetlands. As I mentioned, this may be because there is a strong bias toward swamps in the fossil record; low-lying wet places are the most likely place for a dead organism to be fossilized. But it may also be that forests couldn't make it into the drier uplands despite the innovation of deep roots, fungal symbionts, and myriad water-saving internal structures. Dispersing across long distances on dry soils may have been impossible until plants took one more evolutionary leap—seeds.

Seeds are nutrient-packed, desiccation-tolerant packages that allow plants to disperse long distances, providing young plants a start in life even where soil is dry, nutrient poor, or shaded. Seeds become abundant in the fossil record about ten or twenty million years after *Archaeopteris* dominance, and the appearance of seeds coincides with the collapse of *Archaeopteris* forests. Three hundred sixty million years ago, *Archaeopteris* were replaced by the seed-producing ancient progenitors of modern pines and spruces.

Seed plants had it all—woody structures and fungal partners to help with water and nutrients, and a long-distance travel mechanism that gave their offspring access to a trust fund of the elements in Life's Formula. The transition to land was complete, and from this point forward in Earth's history, plants dominated all but the driest and coldest habitats on Earth. It is unimaginable that their dominance will wane until long after humans have disappeared.

Like cyanobacteria, land plants had two key innovations. First, they found a new way to capture sunlight and thus carbon; in this case the innovation wasn't a new biochemical reaction but the movement of this reaction to a new place. Second, plants got access to nutrients that were stored in rocks. Cyanobacteria evolved a way to get the nutrient nitrogen but remained (and remain) limited by the scarce supply of rock-derived nutrients in the ocean. Land plants could get those nutrients and were able to take their photosynthetic machinery across continents that were virtually devoid of photosynthetic organisms for much of Earth's history. Their innovations in getting water and nutrients allowed their wild proliferation. But, as with cyanobacteria, the story of plants also shows how unprecedented access to life's essential elements can have unintended consequences. Once again, innovation and proliferation ended with catastrophe.

———

To understand what happened next, we have to go back to the interaction between life and the gases that keep the planet warm, the greenhouse gases. After the oxygenation of the atmosphere by cyanobacteria, CO_2 had become the most important greenhouse gas. Our best estimates suggest that the concentration of CO_2 in the atmosphere at the time land plants evolved was about ten times higher than it was at the start of the Industrial Revolution and thus about five times as high as CO_2 will be by the mid-twenty-first century (much more on these modern changes in chapters to come). As I mentioned, those high concentrations of CO_2 created a pan-tropical world, with bathtub-temperature oceans. Back then, snow and ice would have been oddities.

This was the world into which land plants evolved and proliferated. Their success, fueled by access to sunlight and nutrients, coupled with innovations to stay hydrated on land, had global environmental consequences for two reasons. First, as forests grew, they captured CO_2 from the air and stored the carbon in their wood and leaves. To help visualize this, I want to use the analogy of a bank account. It's an analogy I'll come back to several times in this and subsequent chapters. Bank accounts have an amount (the balance), inputs (deposits), and outputs (withdrawals). For carbon, and as we will see in later chapters for other nutrients, there are a few "bank accounts" of importance. Here I'll talk about the air, the land, and the oceans since that's where our story unfolds.

I'll start with the air bank account. The air has a certain amount of carbon in it (mostly in the form of CO_2). At the time land plants evolved, the air had a lot of CO_2—a big balance. As a result, Earth was very warm since CO_2 is a greenhouse gas. But when land plants evolved, they began to suck some of that CO_2 out of the air to build their tissues—a CO_2 withdrawal from the air and deposit into a plant bank account on land. As the world transitioned from having no plants to being a world with forests, the net effect would have been a large withdrawal of CO_2 from the air bank account. That would decrease the amount of CO_2 in the air and begin to cool the planet.

In addition, when plants, their roots, and the fungi and bacteria fed by their roots die, some of their carbon-rich compounds get stuck in the soil. Soil is brown because of the carbon it stores. If you combine all the carbon stored in plants with all the carbon they shunt into the soil (the soil bank account), you get a lot of carbon—about four or five times what is in the air today. Using this carbon accounting, we can see that the transition from a world without plants or real soils to a world with forests and carbon-rich brown soils would dramatically decrease the amount of CO_2 in the air.

There is another piece to the story, and it has to do with the effect plants have on rocks when they mine for soil nutrients. Like all miners, plants don't want everything contained in the rocks they break apart. They are searching for specific things like phosphorus. But even if their mining helps them get what they need, it has consequences. We now know those consequences are also important for the air bank account.

The idea was first postulated by the chemist Harold Urey, who worked on the Manhattan Project. Urey's early experiments tested how life might evolve on a lifeless planet, and he won the Nobel Prize for his discovery of a heavy isotope of hydrogen. I mention him here because he described a series of chemical reactions that we think were key for the way land plants influenced the planet. I've already said that plants and fungi acidify the soil, which speeds up the dissolution of minerals. Urey figured out that the net effect of this reaction further removed CO_2 from the air.

Here's a rough sketch of how this works. The activity of roots and the decomposition of organic material in soil release CO_2 (I'll explain this more in the next chapter; take it on faith for now). This CO_2, in turn, dissolves in the water held in pores in between the soil minerals. That makes the soil water acidic, in the same way that dissolving CO_2 in soda makes soda acidic, and this acid helps dissolve soil minerals. That part we've already covered. What Urey realized is that the CO_2-rich water, along with the other elements dissolved out of the minerals, eventually makes its way to rivers and to the ocean. In the ocean, some of the dissolved CO_2 reacts with calcium, forming calcium carbonate, better known as limestone. The chemical formula of limestone is $CaCO_3$, and the C is carbon. Corals are made of limestone. Limestone is deposited on the sea floor when warm, shallow seawater evaporates. Limestone is the ultimate resting place for the carbon that was once CO_2 in the air.

Urey figured out that, despite this complicated process with both living and nonliving players, the net effect was pretty simple. The dissolving of rocks meant that for every two molecules of CO_2 withdrawn from the air by plants on average one molecule would eventually end up deposited in the ocean as limestone. By speeding up the rate of rock weathering on land, plants accelerated the pace at which CO_2 was transferred from the air bank account into a geologic account at the bottom of the ocean. This process, by which the weathering of rock on land ends up creating carbon-rich rocks beneath the ocean, is a primary regulator of Earth's climate over million-year or more timescales. It is the dominant way that CO_2 is (quasi)permanently removed from the air. Anything that increases the global rate at which continental rocks dissolve will, very slowly, reduce the amount of CO_2 in the air.

Ocean critters can't increase the rate at which rocks weather. They make their living floating on the surface of the water, a long way from any rocks. But land plants can. To return to the theme of this chapter and to the elements that allowed land plants to change the world: land plants' impact was facilitated by two key evolutionary innovations. First, their roots and associated fungi allowed them to stay hydrated on dry land—they had solved the problem of getting H and O. Second, by breaking down rocks replete with other nutrients like phosphorus (P), they got access to the elements that stymied (and still stymie) photosynthetic organisms in the ocean. Rather than depending on rivers and dust for a slow trickle of these nutrients to come to them, plants went to the source. They became the world's best miners. Together these innovations allowed them to spread across the land, withdrawing carbon from the air. This carbon got stored in their bodies, in the soils, and even on the ocean floor as the end product of the accelerated weathering of rocks by plants as they spread across the continents. But this success came with consequences. Over the next hundred million years, the slow transfer of carbon from the air to other accounts (on land and on the ocean floor) began, inexorably, to cool the planet.

As with the cyanobacteria, the final blow was provided by geologic happenstance and the slow movement of the continental landmasses as they were pushed and pulled along with the tectonic plates on which they sat. Around three hundred million years ago, the movement of Earth's tectonic plates happened to produce a lot of swampy lowlands near the equator, where plants were bathed in abundant sunlight. The thing about swamps is that they are great for slowing the rate at which things decompose. Decomposition requires oxygen (again, I'll come back to this in the next chapter), and the thick, organic ooze underneath swamps doesn't let a lot of oxygen in. When plants living in these vast continental lowlands died, even more of the carbon in their tissues was preserved, increasing the net transfer from the air to the land. Another bit of happenstance was that the sea level seems to have fluctuated quite a bit at this time, and the ins and outs of the oceans periodically flooded these coastal swamps, slowing decomposition even more.

Let's sum up. When plants evolved, they pulled CO_2 from the air to build their tissues, and when those tissues died, some of that carbon got stuck in soils. Withdrawal #1 from the bank account of carbon in the air. Plants also accelerated the dissolution of minerals on land, which had the net effect of removing CO_2 from the air and storing it on the ocean floor as limestone. Withdrawal #2. Finally, geologic conditions allowed the growth and repeated flooding of the vast lowland swamp forests. They emerged during what is, not coincidentally, known as the Carboniferous Period. When plants growing in those swamps died, they were protected from decomposition, and their burial over millions of years represented yet another net transfer of CO_2 out of the air. Withdrawal #3. All else being equal, you can't increase the rate at which you withdraw from a bank account without having that account go down. With the triple whammy of withdrawals that land plants imposed, the amount of CO_2 in the air fell precipitously.

I want to draw an explicit parallel here to the events in the previous chapter. If you remember, more efficient photosynthesis was good for cyanobacteria because it allowed them to gather more sunlight. The oxygen produced was just a waste product. Over time, and when the geologic conditions were right, however, this waste product accumulated and caused unprecedented global environmental change. Similarly, sucking up CO_2 allowed plants to grow. Dissolving rocks on land allowed them to gather more nutrients and to further proliferate. The removal of CO_2 from the air and the cooling that resulted from this accelerated weathering and from the burial of dead plants was not beneficial to the plants. In fact, photosynthesis is harder when the air has less CO_2 and is slower at colder temperatures. These changes were just a side effect. At first it didn't matter. For millions of years, the consequences of these side effects were imperceptible.

Eventually, though, plants' innovations pulled enough CO_2 out of the air that the greenhouse effect began to weaken. The pan-tropical Earth, which had supported great forests across most of its land, began to really cool down. It is unclear how long the process took before Earth cooled enough to have ice ages. But by three hundred million years ago, roughly a hundred million years after plants got going in earnest on land, Earth had cooled enough that the vast tropical forests were gone from most of the planet. They were frozen by their own success. Another environmental disaster spurred by innovation, proliferation, and collateral consequences.

The process was slow: a drip, drip, drip out of the bank account of CO_2 in the air and a transfer of that carbon below ground. Some of that carbon was gradually compressed, concentrated, and turned into coal. Then, three hundred million years after those tropical trees succumbed to environmental changes of their own making, the next world-changing organism, humans, discovered that carbon-rich bank account. We began to transfer some of that ancient, stored carbon, originally captured by our world-changing predecessors, back to the air at unprecedented rates.

PART II

Are We So Different?

3

Life's Battery and Earth's Blanket

CYANOBACTERIA AND PLANTS evolved innovations for gathering carbon and using it to store that chemical energy in their tissues. They evolved innovations for acquiring and using nutrients—nitrogen in the case of cyanobacteria and phosphorus in the case of land plants. They (cyanobacteria) used water in new ways, and they (plants) got it in new ways from new places. And in both cases, these evolutionary innovations had cataclysmic environmental consequences. After this sprint through four billion years of life on Earth, we are ready for world-changers 3.0—humans. For the rest of the book, I'll explore what we share, and don't share, with our world-changing predecessors. But the bottom line is this: we are deeply, elementally, connected to our world-changing predecessors. We are enjoying the fruits of our innovation but are beginning to come up against its consequences.

This chapter will tell the stories of humans, carbon, and energy. We'll move on to the way our species interacts with the other elements in Life's Formula in subsequent chapters. There isn't really a perfect place to start, so I'll start with a story about my dear family friend, Sheldon Scott. Sheldon was my parents' age, and when I was a kid, he worked in the woods, running a grader for a lumber company in New Brunswick, Canada. He was one of the smartest people I've ever met and, despite never making it past elementary school, was always asking me questions I didn't know the answers to. As I got older and went further in my schooling, he increasingly loved trying to stump me, which was frustratingly easy for him. When I started "getting crazy about all this environment stuff," he asked me why I cared about CO_2 in the air. "Don't we breathe out carbon dioxide?" he asked. "What the hell difference does it make if it comes out of us or the tailpipe of your car?" It's a good question, and one I'll come back to several times in this chapter. It turns out that the answer is critically important to understanding why human changes to Earth's carbon "bank accounts" and climate are so unusual.

To walk through this question, I'm going to go back to Priestley's bell jar experiment. One of Priestley's great realizations was that Earth is essentially a giant bell jar floating in space. It's a closed system. Very little comes in or out. But that doesn't mean the system is static. There is a lot of movement of material from one place to another. Since we're focused on carbon here, let's go back to our carbon bank accounts and talk about where they are and how much is in them.

Like almost every element on Earth, nearly all the world's carbon is found in rocks. Limestone, which I've mentioned, and marble (which is chemically identical to limestone) have a lot of carbon in them, but many other rocks have carbon in them as well. Nevertheless, the carbon in rocks is essentially irrelevant over human timescales, because it stays put, bound in mineral form and buried deeply away from any living thing. At least it did, until humans got into the game and started digging up and burning carbon-rich rocks like coal. But I'll come back to humans in a few pages. First, I want to set the stage for how the world was working before we started to "exert all [our] powers" at their current level.

For all of Earth's history until the Industrial Revolution, the only way carbon escaped from deeply buried rocks was when CO_2 erupted out of volcanoes. While a single volcanic eruption is dramatic, on average, over years, centuries, or millennia, the amount of CO_2 escaping from volcanoes remains pretty constant, even if there is a big eruption somewhere in a particular year. Thus, volcanoes are part of what I'll term the "slow carbon cycle." They make deposits of carbon into the air account. In the last chapter, I talked about how land plants, the carbon they stored in their tissues and soils, and the weathering of rocks all represented net withdrawals of carbon from the air. These withdrawals were also part of the slow carbon cycle. Cumulatively, over millions of years, the proliferation of land plants pulled enough CO_2 out of circulation to plunge a once pan-tropical world into a large-scale glaciation.

In addition to this slow cycling, the carbon that escapes from rocks into the air via volcanic eruptions circulates around Earth's surface very rapidly. Photosynthesis captures roughly as much carbon from the air in a day as volcanoes emit in a year. But the vast majority of this "fast"-cycling carbon is shunted back into the air through a process called respiration. I've talked about respiration before, but now I want to spell it out chemically—it is the reverse of photosynthesis:

CO_2 + H_2O + sunlight energy → sugar + oxygen
(photosynthesis)

sugar + oxygen → CO_2 + H_2O + usable energy   (respiration)

Photosynthesis captures the sun's energy and stores it in chemical form (as sugar or some other biological material). Respiration occurs when I, or a zebra, or a wood-rotting fungus, consumes that sugar (or other organic molecules). We use the released chemical energy to fuel our bodies, and we breathe out CO_2 and water. These two opposite reactions, photosynthesis and respiration, drive the fast carbon cycle.

Over a decade, photosynthesis withdraws enough carbon from the air to empty the air bank account. But as long as the total withdrawal of carbon from the air by photosynthesis is balanced by the total amount deposited back into the air by respiration. But as long as the total withdrawal of carbon from the air by photosynthesis is balanced by the total amount deposited back into the air by respiration, the total amount of carbon in the fast carbon cycle will remain the same. Only volcanoes, the weathering of rocks, and the long-term burial of organic carbon (like plants in swamps) can change that. When I said in the last chapter that the burial of plants led to a long-term reduction in CO_2 in the air, what I meant was that there was slightly more photosynthesis than respiration. Each year, or more likely each millennium or longer, a tiny fraction of the carbon captured by photosynthesis escaped being respired back into the atmosphere. Think of a fishbowl with a tiny, tiny leak. For weeks, months, or even years if the leak is small enough, fish can go about their business and barely notice the falling level of water. But eventually the water gets low enough that fish business as usual is no longer possible. That's how it was with tropical forests three hundred million years ago—eventually the level of CO_2 in the air got so low that the climate got cold enough for glaciation. That was the end of business as usual in what had been a tropical world.

Unless there is a directional change—a consistent imbalance in one way for a long time, the fast carbon cycle is just noise over the longer term. Here's an example of what I mean by noise. Most of Earth's land and therefore most land plants are in the northern hemisphere. The summer bloom of northern plants takes CO_2 out of the atmosphere and stores it in leaves. In fact, in the northern summer, total global photosynthesis outstrips respiration, and CO_2 levels in the air drop. We can measure that drop all over the world. But those leaves die in the fall, and photosynthesis in the northern latitudes slows. Those leaves decompose over the winter, and atmospheric CO_2 levels rise back up to what they were the previous spring. Viewed over a year, this is a strong signal. But viewed over a century, it's just a small fluctuation around the average. The total amount of carbon in circulation stays the same from year to year (before humans existed, anyway). If you measure CO_2 in a forest, the same thing happens over the course of a day. During the day photosynthesis outstrips respiration, and CO_2 levels in the forest fall. At night respiration outstrips photosynthesis, and CO_2 levels rise. But the daily average doesn't change until the seasons do. The long-term averages are stable unless something unusual happens, like plants emerging to cover the land in forests, or geologic happenstance creating vast, low-lying, wet continental areas where huge quantities of plant material are buried faster than they can decompose. And even these unusual events take millions of years to unfold. No organism would be able to perceive a change during its lifetime.

In the time since land plants cooled the planet three hundred million years ago, there have been a few massive volcanic eruptions that rapidly injected a lot of CO_2 into the air, changed the climate rapidly, and caused mass extinction events. The last one of these was sixty-five million years ago, when a meteor and massive volcanism doomed the dinosaurs. More "recently" (in the past fifty million years), the total amount of carbon in the air bank account has been gradually decreasing. Some scientists think this decrease is the result of tectonic forces, though there is a lot of debate about the details. We know that over that time Earth's tectonic plates have been pushing the Indian subcontinent into Asia, elevating the massive Himalayan peaks. The high mountains shed fresh minerals exposed in the collision, and those minerals dissolved quickly in the warm, wet foothills and lowlands. Since rock weathering consumes CO_2, the CO_2 content of the air has been falling ever since. As a result, the planet gradually cooled. It got so cool two million years ago that Earth was again cold enough to have ice ages. This is another leaky fishbowl scenario. Over the past fifty million years, Earth has gone from one of its warmest periods to one of its coolest, driven by a drip, drip, drip drawdown of CO_2.

It's important to pause here and think a little about the vastness of geologic time. My wife Beth, who was a geology major in college, often says that through geology she came to viscerally understand the difference between a billion and a million. A billion years is a long time in geology, but a million years is not. When I say CO_2 levels fell over the past fifty million years, I mean that they fell very, very slowly for living creatures. But they were actually falling pretty fast relative to other periods in geologic time. Our best guess is that the concentration of CO_2 in the atmosphere fell sixfold over the past fifty million years (from about 1,200 parts per million to about 200 parts per million; parts per million [ppm] is a measure of concentration). That's a decrease of 20 ppm *every million years*. That decline took Earth from its warmest point, when there was no ice at either pole and ferns growing in Antarctica, to one of its coldest. That's rapid change for the slow carbon cycle. But when humans discovered fossil fuels, "rapid change" to the carbon cycle took on a whole different meaning.

———

The earliest record of coal use dates back almost six thousand years in China, and in nearly every place coal was easily found, it has been used by humans as a source of heat and energy. The first modern commercial coal mine opened in Virginia in 1748. Most of the deposits that were first exploited, in England, northern Europe, and the United States, were the remnants of the tropical forests and swamps that were the focus of the last chapter. These deposits formed as plants sucked CO_2 out of the air and got buried slightly faster than they could be consumed and respired. But we are consuming them now. Instead of eating them—metabolizing their energy and breathing out CO_2—we burn them. First in steam engines, factories, and houses. Now in power plants and in industrial facilities. It has been estimated that we are currently burning the remains of more than four hundred years of buried photosynthesis every year [5].

This brings us back, finally, to my friend Sheldon's point. If you look at the chemistry of fossil fuel combustion and respiration, the two processes seem remarkably similar. Respiration takes sugars or other organic molecules and breaks them down to release metabolic energy, CO_2, and water. Gasoline combustion combines an organic molecule (octane, for example, is eight carbon atoms linked to eighteen hydrogens) with oxygen to produce carbon dioxide, water, and, in the process, energy to power your car.

$$\text{sugar} + \text{oxygen} \rightarrow CO_2 + H_2O + \text{usable energy} \quad \text{(respiration)}$$

$$\text{gasoline} + \text{oxygen} \rightarrow CO_2 + H_2O + \text{usable energy}$$
$$\text{(combustion)}$$

In this way Sheldon was right. But there are a few differences that turn out to be important. First, respiration uses sugar (or starch, or protein), and combustion uses gasoline (or coal or natural gas). But fossil fuels are just organic molecules, compressed, stored, and tweaked a little over geologic time. In fact,

FIGURE 5. This page, a coal mine in Illinois with a ceiling full of fossil plants; opposite, a close-up of one of our world-changing predecessors, this one about 310 million years old. Burning coal releases CO_2 through essentially the same chemical reaction that occurs when you eat a salad and your digestion produces CO_2. The difference is that the coal contains carbon that has been separated from the fast carbon cycle for 300 million years, whereas your salad is made from carbon that was or is already moving freely around Earth's surface. Thus, burning coal increases the total amount of carbon in circulation at Earth's surface (moving between air, land, and water), and eating salad does not. Photo courtesy of Dr. William A. DiMichele, Smithsonian Institution.

in many coal deposits, you can still see the fossil fern leaves whence the coal came (Figure 5). So that difference isn't so big after all. Second, respiration produces energy inside the body, and combustion makes energy outside the body. But if you're clever enough to build a fireplace, a car, or a power plant, that combustion energy can be used anyway. That's a key difference, since we use energy for all sorts of things and not just metabo-

FIGURE 5. (*continued*)

lism. I'll come back to this idea later in the book; it will be very important. Third—and this is key to answering Sheldon's question about the difference between burning gasoline and breathing out during respiration—is that respiration uses carbon that was already in the fast carbon cycle. If you eat an apple, that apple has sugars made of carbon that was taken out of the air bank account just a short time ago. If that apple fell before you ate it, it would decompose (essentially get eaten by bacteria) and go back to the air as CO_2 anyway. The net change to the air account would be zero. In contrast, fossil fuel combustion takes carbon that was stuck in the slow cycle, deep underground and out of circulation, and pumps it into the fast cycle, ramping up the total amount of carbon in circulation.

Humans are the first organisms in the history of life who have bridged the slow and fast carbon cycles in a way that increases the total amount of rapidly cycling carbon. The CO_2 we are emitting by burning fossil fuels, unlike the CO_2 we breathe out after eating, contains carbon that was pulled out of the fast carbon cycle three hundred million years ago (at least in the case of most coal). As a result of this bridge, more fast-cycling carbon is becoming available every day, as we release the CO_2 from coal, oil, and natural gas combustion into the air. The extra carbon is rapidly shuttled in the fast carbon cycle as photosynthesis and respiration occur. But there is more and more carbon everywhere. More in the air, more dissolved in the oceans, and more stored on land. To go back to the fishbowl analogy, we are rapidly pouring more water into the tank. Fish business may well continue as normal for a bit, until the tank overflows and the system changes fundamentally.

What happens to CO_2 once it enters the fast carbon cycle is very complicated, and many scientists, including me, are racing to understand all the consequences of our bridging of the fast and slow carbon cycles. How much CO_2 will stay in the air? How much will be taken up by plants or stored in soil? As more CO_2 in the air warms the planet, will carbon that is currently stored in plants and soil decompose and be respired as CO_2, exacerbating warming? Will forests like the Amazon start experiencing long dry seasons so that human-caused fires increasingly produce raging wildfires, which inject carbon that was stored in wood into the air? All these questions and many more have profound implications for the exact trajectory of climate warming in the twenty-first century because these factors will influence how much CO_2 is in the air.

Despite these uncertainties, though, there is no doubt that the amount of carbon in the fast cycle—the carbon shuttled between air, land, and water—is increasing. From the Industrial Revolution, say around 1850, the amount in the air has increased by almost 40 percent as a direct result of our burning of fossil fuels. It will continue to increase until we eliminate fossil fuel combustion and/or figure out a way to capture CO_2 from the air and convert it back into rocks at an almost unimaginable scale. Instead of one fishbowl imagine three, all interconnected. In this case the fishbowls represent the ocean, the land, and the air. If you pour a bunch of water into one bowl, the water will rise in all three. Exactly how much ends up where depends on the intricacies of their connection, but the total amount in each bowl would go up, especially in the one you poured the water into. We are releasing billions of tons of CO_2 into the air each year by burning fossil fuels. So far, a little less than half of what we've emitted has stayed in the air. The rest has been taken up by photosynthesis on land or dissolved into the oceans (which are acidifying as a result—remember that CO_2 dissolved in water is also what makes soda acidic).

To reiterate the answer to Sheldon's question: there is a difference between the carbon dioxide we breathe out and the carbon dioxide that comes out of a tailpipe even though both are CO_2. Eating a plant releases carbon that was already in circulation, carbon taken from the air by plants one time and released as CO_2 back into the air by something that ate those plants a short time later. Burning fossil fuels is different. It is putting new carbon into play. This practice has turned humanity into an ever more active industrial volcano, releasing carbon that had been stored in rocks for eons. Some of that carbon stays in the air, some of it goes into the ocean, and some goes into the land. But the total amount in circulation is going up. Fast.

Just how fast? Land plants, their effects on weathering, and their eventual burial to form coal deposits were the focus of chapter 2. This world-changing may have dropped the concentration of CO_2 in the air from as high as 4,000 ppm (that's the highest of many different estimates) to something like 400 ppm in the span of one hundred million years. That's a very rapid geologic rate—a 34 ppm drop per million years. In the last fifty million ppm drop per million years. That rate, tens of ppm per million years, is "rapid" in the geologic sense. Both took Earth from a very warm period to a very cold one and changed life dramatically for virtually every organism on the planet.

In contrast, in the last two hundred years, humans have increased the carbon dioxide concentration of the atmosphere from 280 ppm to 416 ppm (as of June 2020), and the concentration since 1961 is climbing by about 4 ppm per year (Figure 6). I won't ask you to take out your calculator—we're releasing carbon from rocks about one hundred thousand times faster than those "rapid" geologic changes. Put another way, in each recent decade humans put as much extra CO_2 into the air as the land plants pulled out in a million years when they changed the world.

As with cyanobacteria and plants before us, our changes to the amount of CO_2 are collateral damage. We wanted energy and innovated to get it from a geologic bank of organic carbon. We combined that banked carbon with oxygen (that's what burning is), which liberated that ancient, stored sunlight energy for our own purposes. It just so happens that the process emits CO_2—a waste product, and one with consequences. Just like our world-changing predecessors, we benefited tremendously from our innovation. We used it to lift billions of people out of poverty. Life expectancies have soared. The percentage of malnourished humans has plummeted. Our population has boomed, growing from one billion in 1800 to two billion in 1927, to three billion in 1960, to almost eight billion today. None of that could have happened without a new source of energy. But our rapid

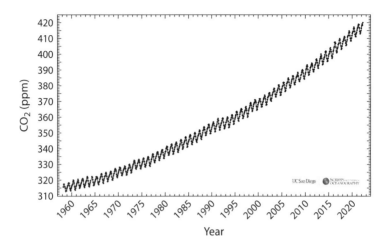

FIGURE 6. There are two signals in this graph. The first is the steep rise in CO_2 concentrations in the air, resulting mostly from fossil fuel combustion by modern society—linking the fast and slow carbon cycles. The second signal, the sawtooth pattern on that rising curve, is driven by the spring/summer bloom of leaves in the northern hemisphere, which pulls CO_2 from the air, and the fall and decay of those leaves in the northern fall/winter, which release that CO_2 back into the air. But the upward slope from burning fossil fuels dwarfs the sawtooth signal of the land plants. Figure courtesy of the Scripps Institute of Oceanography.

proliferation enabled by innovation and the release of waste products that come with it can have world-changing side effects.

Having built an understanding of the way carbon flows around our planet and how humans are changing those flows, I want to turn now to the consequences of those changes. Once again, I will try to give some context by looking into the past, but this time tens of thousands, not millions or billions, of years ago. As with other world-changers, the consequences of our changes to the air are playing out through climate change. Since humans are changing the concentration of greenhouse gases in the air at unprecedented rates, the global climate is responding with similar rapidity. I want to take a couple of pages to talk about just how fast temperatures are changing.

The long-term global cooling over the past fifty million years eventually removed enough CO_2 from the air to transition the world from a global hothouse to a global icehouse—one in which tiny wobbles in Earth's orbit are sufficient to trigger ice ages. Over the past two million years, Earth has come in and out of ice ages as the wobbles have initiated chain reactions that moved CO_2 from the air into the land and oceans (causing cooling and ice ages) or from the land and oceans to the air (causing warming and ending ice ages). The last ice age ended about ten thousand years ago, and since then Earth's average temperature has warmed by about 7°F.

Seven degrees Fahrenheit is a lot—enough to move Earth from an ice age to the relatively warm conditions Earth is in now. To give you a sense: during the last ice age, continental ice sheets reached all the way from the North Pole to where New York City is today. You can see the glacial scars in the bedrock exposed in Central Park, topped by erratics—giant boulders transported from the Adirondacks on a slow-moving conveyor belt of ice. The continental ice sheets stored so much water that the sea level was 125 meters (410 feet) lower than it is today. If it weren't for the massive ice sheet, you could have walked across Long Island Sound from Connecticut to Long Island. The coast was over 100 miles farther east, way out in what is now the Atlantic Ocean.

Why does such a seemingly small change make such a big difference? After all, the temperature changes more than 7°F almost every day. How can the same change in Earth's average temperature cause ice sheets to advance as far south as New York City, as it did ten thousand years ago, and then melt back so there aren't ice sheets in North America anymore?

The answer is somewhat counterintuitive, but while a seven-degree change over a day or between seasons isn't a big deal, a seven-degree change in the long-term average temperature of Earth represents a dramatically different climate. As an example, in 1815, a volcano in the South Pacific called Mount Tambora erupted, spewing so much ash into the air that it blocked out the sun. Crops around the world failed, starvation ensued, and 1816 was a European "year without summer" [30]. How much did the average temperature of the planet go down as a result of that eruption? About 1°F. A small shift in a global average is a big deal.

Let's probe a bit more into why this is the case by looking at a non-climate example. Imagine you measured the height of all the people on Earth. You would come up with an average height of about 5′4″ (162 cm). What would it take to shift that average? Not the birth of a few more very tall or very short people. It would take a global shift in childhood nutrition and access to health care, big changes to the whole system. Indeed, heights have risen globally in the past couple of centuries for exactly this reason. Similarly, for Earth's average temperature to go up, the average of every place on the planet from the poles to the equator, in effect, the whole system, must experience a profound shift in the same warmer direction. Shifting the average of lots and lots of numbers requires a big push.

Another reason that small shifts in the average are a big deal for the planet has to do with the way changes in the average affect the frequency of extreme events. To understand the point, let's imagine a city not unlike my hometown of Providence, Rhode Island, where the average low temperature in January is 22°F (~5.5°C). That's about what it was when I was a kid in the 1970s. But Rhode Island winters have warmed by about 4°F since then, so the average lately has been about 26°F

(−3.3°C). Now the difference between 22 and 26°F doesn't seem like a big deal. Either temperature will require a hat, gloves, and a warm coat if you're going to be outside for any period of time. But just because you can't feel a difference in the average doesn't mean it doesn't matter.

Below, I've plotted two identical bell curves (Figure 7). One has an average of 22°F (the average winter low in Providence in the 1970s, shown in black dots), and the other has an average of 26°F (the average winter low in Providence now, shown in gray dots). The graph illustrates how common it is for the low temperature of the day to be a particular temperature. For example, if you're looking at the black line, you'll see it was very common for the low to be between 15 and 25°F but rare for the low to be below 10°F or above 35°F.

Now look at the right-hand side of the graph and the dashed line I've drawn at 32°F, the freezing point of water. When the average was 22°F, it was very rare for the low temperature to get hotter than 32°F. As a result, most of the precipitation fell as snow, and there was snow on the ground most of the time. But with a warming of only a few degrees, it has become common for the low temperature to be above 32°F. As a result, it's common to get wintertime rain, and there is rarely snow on the ground.

I show the curves to make a point, one that people all over the world are learning through experience. If you look up the actual data for Providence (I did), you'll find the number of anomalously hot January nights has risen from thirteen per decade to thirty-four since the start of the twentieth century. But you don't need to dive into the archives. Ask anyone over the age of fifty—at least anyone from a cold, snowy place—if the winter is the same as it used to be, and the answer will be no.

All this is to make the point that a few degrees change in the average temperature across all of Earth's surface means things have really changed a lot. In this context, the 7°F warming at the

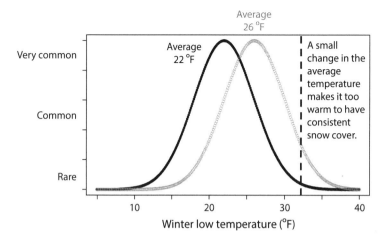

FIGURE 7. A small change in the average temperature in Providence, Rhode Island, has a big effect on the extremes.

end of the last ice age was huge, even if it took thousands of years to occur. It's the difference between a mile of ice over most of Canada and the northern United States ten thousand years ago, and no permanent ice in the United States today (except on the tallest mountains).

With that context, let's look at the human effects on climate. We've already caused almost 2°F of warming of the planetary average since 1900, most of which has happened in the last fifty years. That's pretty fast. By the end of this century, without dramatic reductions in our inadvertent manipulation of the global carbon cycle, we could pump enough greenhouse gas into the air to raise the global average temperature by 7°F. That's as much as the increase between the end of the last ice age and the start of the Industrial Revolution. If we follow that path, we will have reversed almost one million years of cooling in the span of less than two centuries.

It would take centuries for the full effects of such an increase to be felt, but that future planet would be unrecognizable. My university, which describes itself as a college on a hill, would at best be a college on an island. New York, Singapore, Tokyo, Beijing, Mumbai, and many island nations would be swallowed by the sea. On land, most of the places we currently grow food would no longer support agriculture, at least with the crops we know how to grow today. Heat waves that were once-in-a-century occurrences would be the norm, and many places that are already hot today would be too hot to sustain human activities. This cataclysm would take just a few human lifetimes to unfold. Long to us, but the blink of a geologic eye. Humans may well survive a transition to this pan-tropical world. But they also may not. Living through changes that result from our kind of innovation, innovation rooted in access to life's key elements, is neither pleasant nor guaranteed.

As depressing as this is for a whole host of reasons, it's also kind of amazing. No animal in the history of life on Earth has been such a dominant force of nature that their imprint on the planet was necessary to understand the climate. Even bacteria and plants, which are far more important for the climate than animals, are incapable of changing the planet over a hundred years. A hundred million? Sure. But a hundred? No way.

Yet in another way we are not so unique. We've simply done what has been done before. We found a new way to get energy from carbon and are exploiting it for all we're worth. Indeed, based on the lessons we've learned, the question is not whether humans are capable of changing the planet. The question is how all this fossil fuel use could *not* change the climate? We have known for almost two centuries, since the pioneering work of Eunice Foote in 1856, that greenhouse gases trap heat. If you calculate how much additional greenhouse gas there is in the air as a result of human activity, you can figure out how much extra heat is being trapped because of that gas. The amount is staggering— it's equivalent to eight times the energy of the bomb that was

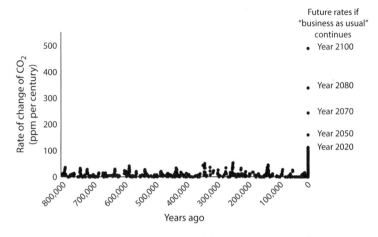

FIGURE 8. The best record we have of the concentration of CO_2 in the air comes from ice cores in Antarctica. By drilling down into the ice and liberating tiny bubbles of air trapped in the ice as it was deposited year after year, we can directly measure the concentration of CO_2 for the past eight hundred thousand years. The difference between the warm, stable climate that allowed human civilizations to rise in the past ten thousand years and a full ice age with glaciers covering New York City is the difference between concentrations of CO_2 in the air of 280 ppm (warm) and 180 ppm (cold). Today, CO_2 concentrations are about 415 ppm. To give a sense of the rapidity of human changes, oceanographer Jim Barry put the ice core data together with modern measurements and calculated the rate of change in CO_2 concentrations from century to century. During that time, Earth swung "rapidly" from an ice age to a non–ice age nine times. But humans have redefined rapid change, and those rapid cycles are dwarfed by the changes over the late twentieth century, not to mention those that await us without a change in our use of fossil fuels.

dropped on Hiroshima exploding every second of every day. How could we add that much energy to the surface of the planet and not warm it up? Changes to the carbon cycle, especially in the amount of carbon in circulation, always change the planet. As far as we know, it just hasn't ever changed it this fast (Figure 8).

Given that the primary driver of climate change is our emission of greenhouse gases, our bridging between the fast and slow carbon cycles by the burning of fossil fuels, it is not surprising that the biggest variable moving forward is what we choose to do. If we rapidly transition from carbon-based energy, an option available to us but not to our cyanobacterial or plant predecessors, Earth will continue to warm, but it will remain fairly recognizable. If, however, we continue as we have been, the world really won't be recognizable. It's a bleak picture that has been painted many times. But it is not destiny. Unlike land plants and the cyanobacteria before them, we do not have to do business this way. Our world-altering chemistry gives us a way out.

To explain what I mean by this, I'll go back to the three chemical equations I offered at the start of this chapter for photosynthesis, respiration, and gasoline combustion. The first two have been the keys to life on Earth for almost four billion years. Some organisms capture CO_2 to build their bodies; others eat them and use that stored energy for themselves. Humans do the latter—we eat, break down the bodies of other things (plants and animals), and breathe out CO_2 and water. But we've also figured out a way to use energy that was produced outside our bodies. First, we used fire (which has the exact same formula as respiration), and ultimately, we stumbled upon fossil fuels, which provide us energy in very much the same way.

For all this energy we consume from combustion—to run our cars, heat our houses, produce our electricity, and run our manufacturing—our need is energy, not carbon. I mentioned that difference a few pages ago, and I want to come back to it now. It is another key difference between eating food and combusting gasoline. We need the chemical energy in food to build our bodies and to keep our cells running. We can't replace that carbon; like all organisms that consume other things, we're stuck with that chemistry. But in the energy we use from outside our bodies, we have no need of carbon—it's just an energy carrier. It can be replaced. It's already being replaced by solar panels and windmills and hydroelectric dams and nuclear power plants. We've figured out how to get energy in other ways. This is our way out, because there are a lot of ways to get energy that don't need carbon.

Way more solar energy hits Earth in an hour than is consumed in a year of burning fossil fuels. That sunlight can be transformed, either directly or indirectly, into electricity that can be readily transported to where energy is needed. Winds, driven by the differential heating of different parts of Earth, also provide a source of energy that dwarfs our global energy consumption. Nuclear fuels, while not without serious environmental concerns about their use, can provide a source of energy largely decoupled from either the fast or slow carbon cycle.

Together these alternatives provide an obvious path towards avoiding the climatic cliff both cyanobacteria and plants inadvertently tumbled off. Unlike them, we can see what's coming. But there is no reason to think we won't go off a similar climatic cliff if we refuse to change how we get our energy. Unlike our world-changing predecessors, we can make the switch. The carbon-based energy in fossil fuels is finite, and we have to stop using it long before it runs out in order to avoid climatic catastrophe. Luckily, that energy is replaceable. As we'll see in the following chapters, the rest of the elements in Life's Formula are not.

Even if fossil fuels are replaceable, the switch from them will not be easy. It would represent something that has never happened before—a global move in the short-term to avoid long-term consequences. It will be costly, difficult, and undoubtedly accompanied by consequences we don't yet anticipate. Of course, the same is true for climate change. But it's still going to be a big shift from an energy source that has brought human-kind unprecedented prosperity. I think it's worth acknowledging that, and for those of us calling for a rapid end to fossil fuels (very much including me) to be really sure that the alternative is better. So, before turning to how humans use the other elements in Life's Formula, I'm going to break from the narrative of evolutionary innovation by world-changing organisms. The next chapter will look a little more deeply into how we know what the future might look like if we do or don't rapidly ween ourselves off our use of fossil carbon. Specifically, how can we be so sure that our changes to the fast carbon cycle will, if left unabated, precipitate catastrophic climate change? I don't want you to have to take such a grave pronouncement on faith.

4

How We Know What We Know
about Climate Change

MY MOTIVATION in taking a detour to explain a bit about how
we know humans are causing climate change, and how we can
use that knowledge to predict the future, stems from my teaching at Brown University. I think it's fair to say that Brown is a
very liberal institution in a very liberal part of the United States.
As such, almost all of its students and faculty ardently believe
that humans are causing climate change, just as other people
ardently do not believe this. But whether or not humans are
causing climate change and whether we know enough about
human changes to the climate to make predictions about the
future is not a matter of belief. It is a matter of evidence. Yet this
evidence is rarely laid out in plain language, and my experience
is that few people, regardless of their beliefs, really understand
it. As a result, I find that even some of my most fervent climate-
activist students cannot articulate the broad brushstrokes of
how we know what we know. Given the importance of the topic
to the future of humans and all other organisms on the planet,
I think it's worth the time to digress a bit into how climate
change science works.

Let's start with the statement "Humans are causing climate change by burning fossil fuels." This is essentially the statement I ended the last chapter with. It is the basis upon which people all over the world, including me, are calling for the rapid end to fossil fuel use and the transition to carbon-emission-free energy sources. It's a pretty bold statement, and it is very different from saying *that* climate is changing—what scientists call "detection." If we're going to argue for a massive change in human society, which is what will be required to end our use of fossil fuels, it seems reasonable to ask that we move beyond detection. After all, fossil fuels, despite their problems, have provided tremendous benefits to society over the twentieth century. If we (the climate-concerned public) are going to insist that we stop using fossil fuels, it is incumbent upon us to prove that the downside is greater than the very real upside that fossil fuels have offered. We need to prove, beyond a reasonable doubt, that climate is changing *and* that human use of fossil fuels, not something else, is responsible for the climate change we are observing. We need attribution in addition to detection. Attribution of climate change to human emissions of greenhouse gases is the first claim I'm going to work through in this chapter.

The second claim builds on the first. For attribution, we need a good understanding of how the climate works and what controls it. Once we have that, we can use what we know to make predictions about what the future may hold if we change those controls. This allows us insight our world-changing predecessors never had. We can see what's coming. But let's start with claim number one: human emissions of greenhouse gases are causing the climate change we see today.

For the purposes of this chapter, let's define climate as the long-term (decadal) average of weather in a place. In the last chapter, I talked about winter low temperatures. That's part of a description of the climate of a place. So is saying that Riyadh, Saudi Arabia, has hot, dry summers or that Rio de Janeiro, Brazil, has a warmer climate than Calgary, Canada. It doesn't matter if a particular summer day or week in Riyadh gets chilly or if on a particular August day Rio is colder than Calgary. What happens on a given day is weather. The average of lots and lots of weather is climate.

In this context, climate science is about explaining long-term patterns in temperature and rainfall and changes to them. I like to think of science as a detective story, and climate science definitely fits the bill. In this case, we've found evidence of a crime—climate is changing around the world. The twenty warmest years since the late 1800s occurred in the last twenty-two years. The 2010s were the decade with the fastest warming, and the 2000s were right behind them, followed by the 1990s and the 1980s. Beyond the data taken directly from thermometers, there have been tens of thousands of individual studies that explore how climate is changing in a particular place. They pretty much all point in the same direction—it's getting warmer. Warming is causing ice melt in the Himalayas and threatening the water supply of almost a billion people. Almost all glaciers around the world are retreating at an ever-accelerating pace. Spring thaw is coming earlier, and fall frost later, in every place that regularly gets either.

More subjectively, ask anyone over the age of fifty if winters are the same as those when they were kids, and they will say no. Having grown up skiing in Vermont and having just turned fifty, I can add my voice to this chorus. In Alaska and Siberia, ice roads used to service oil fields and mines are becoming increasingly impassable and dangerous as the ice and the permafrost beneath them melts. On the other side of the planet, the Great Barrier Reef, the largest living structure visible from space, is dying in rapidly warming water. Scientists are not supposed to argue by anecdote, but the accumulation of tens of thousands of anecdotes is compelling. For me, the most compelling climate change anecdote is that "Jingle Bells," the classic Christmas song, was written in 1850 as a song for Thanksgiving sleigh races in Massachusetts. I grew up nearby, and I can't remember a Thanksgiving in my lifetime when we could have counted on sleigh races in Massachusetts.

OK—the world is warming. There is a mystery to solve. Now we have to figure out "who done it?"

———

How does a detective solve a crime? How do they go through the list of suspects, eliminating some and homing in on others, until they find the culprit? At least on TV, they interview the suspects. They build up a good understanding of all the players, how they act, what they did, when they did it, why, and how. Then they try to reconstruct the crime and figure out who did it. If the accused doesn't confess, the case goes to trial by jury. The jury is presented with the facts—the reasoning about who did what and when. The jury is asked to determine their conclusion "beyond a reasonable doubt."

Climate science works the same way. Since we know greenhouse gases play a big role in driving climate and we know human combustion of fossil fuels has increased the concentration of greenhouse gases in the atmosphere by almost 40 percent (and deforestation and agriculture have added more), we have a reasonable suspect. We might even go so far as to say we have strong circumstantial evidence of the suspect's guilt. But like good detectives, we need to make sure we have causation— evidence that our emissions of greenhouse gases are causing warming. We can't be satisfied with the circumstantial evidence of correlation. Someone's presence at the scene of the crime doesn't make them the criminal. We have to rule out coincidence if we are going to be convinced beyond a reasonable doubt. How could we even go about that?

It would be nice if we could use a traditional experimental setup, as you might in a high school or college science class. Imagine you want to know if a particular antibiotic kills a particular strain of bacteria. You could grow the bacteria, perhaps ten batches of identical bacteria, all in their own flasks, in a lab. Then you could give five of those flasks a treatment of liquid antibiotic and give the other five flasks (called the "control") the same amount of liquid but without the antibiotic. If the five flasks that received antibiotics have no bacteria left alive after the treatment, but the control flasks were still brimming with living bacteria, you'd be pretty sure that the antibiotic kills the bacteria. Maybe the person at the next table in the lab sets up a similar experiment but uses different flasks, a different way of putting the antibiotics in the flasks, or a different concentration of antibiotics. If they got the same result, you'd be even more sure that the antibiotics are effective at killing those bacteria.

There are two key components to this experiment. First, the only difference between the treatment group and the control group is the addition of antibiotics. Second, there are five (not just one) flasks in each group. This replication makes it less likely that the bacteria in the five treatment flasks all spontaneously dropped dead for some reason unrelated to the treatment while those in the five control flasks just happened to avoid a similar fate. It is much more likely that the antibiotic treatment was the difference. If there had been ten flasks getting antibiotic and ten controls, and we got the same pattern, we could be even more sure that the outcome was not solely the result of chance.

This classic experimental design is a real problem for those studying climate change. Humans are changing all sorts of things about the planet, all at once. In addition, Earth's climate changes naturally on its own. The Pacific Ocean sloshes slowly back and forth a couple of times per decade, causing El Niño/ La Niña climatic shifts. The sun's energy output waxes and wanes over the decades. Even worse, only one Earth exists. We don't have five replicated Earths getting the human treatment and five control Earths without people. Instead, we are running an uncontrolled, unreplicated experiment on the only Earth we have. How can we be sure of what is causing what?

The answer is that we build simulated Earths in a computer and do experiments on them. My colleague Baylor Fox-Kemper, an oceanographer and climate scientist, recently summarized the approach in an interview with me for a course on climate solutions. "Climate activists like to say, 'There Is No Planet B,' but that's exactly what we create in climate models. Inside a computer, you can have as many planets as you want." At first blush, I'm betting this isn't overwhelmingly compelling. Computers are just powerful calculators. How can one do an experiment in a calculator? And should we believe the results of such an experiment? What I hope to convince you of by the end of this chapter is that we can, and we should. It's the only way forward in this detective story.

Doing a climate experiment in a computer depends on understanding how Earth's climate works, at least well enough to mathematically represent the relevant processes. People have been doing this for longer than you might think. As long ago as 1896, Svante Arrhenius, a Swedish chemist, created a climate model to investigate the causes of ice ages. In a paper titled "On the Influence of Carbonic Acid [as CO_2 was then called] in the Air upon the Temperature of the Ground," he laid out a detailed understanding of the short- and long-term carbon cycles that we covered in the last chapter. Arrhenius noted, "The most important of all processes by means of which carbonic acid [CO_2] has been removed from the atmosphere in all times, namely the chemical weathering of siliceous minerals [rocks], is of the same order of magnitude as a process of contrary effect, which is caused by the industrial development of our time [burning coal], and which must be conceived of as being of a temporary nature" [6]. In other words, burning fossil fuels will increase the amount of CO_2 in the air, but fossil fuels will eventually be used up.

In the same paper, Arrhenius walks through some other remarkably prescient calculations, noting, "I should certainly not have undertaken these tedious calculations if an extraordinary interest had not been connected with them." He calculated that if the CO_2 concentration in the air fell below about 200 ppm, it would be cold enough that continental glaciers, such as existed in the last ice age, would spread across the land. A century later, ice cores from Antarctica revealed that just below 200 ppm was almost exactly the concentration of CO_2 in the air during the last ice age. Arrhenius was actually so concerned about the return of an ice age that he suggests that burning a lot of coal would be a good thing since it would forestall the otherwise likely readvance of the ice. He predicted that doubling the amount of CO_2 in the air would lead to a 5 to 7°F rise in the average surface temperature of Earth, with roughly twice that amount of warming in the arctic, and more warming at night and in the winter than during the day and in the summer.

We are not yet at a point when the concentration of CO_2 in the air has doubled over what it was when Arrhenius was writing.

But he also calculated how much Earth would warm if CO_2 levels went up by 50 percent, which is roughly where we are now, and his numbers are remarkably in line with the changes we are currently observing. Not only that, but his predictions of where (the poles more than the equator) and when (winters and nights more than summers and days) warming would be most pronounced were dead on.

Arrhenius was able to be so accurate, even with very little information, because the temperature of a planet's surface depends on just a few things. First, it depends on the energy absorbed by the surface—the amount of sunlight coming in minus however much is reflected by the surface. This incoming energy will warm the surface, which in turn will radiate energy. The surface will warm up until the amount of energy absorbed is compensated by the amount of energy radiated away. All other things being equal, if you put an object near a warm fireplace, it will get warmer than the same object sitting farther away. That's because if it's closer to the fire, it will warm up until the amount of energy it is radiating is the same as the amount it's absorbing, and warmer objects radiate more energy. The exact same thing is true for planets.

It turns out that this simple energy balance idea is pretty much all you need to calculate the surface temperature of Mercury, Mars, or the moon. But for Earth, we need to understand one more thing—the air. As I've discussed at length, the greenhouse gases in the air—carbon dioxide, methane, nitrous oxide, and water vapor—are transparent to visible light, so sunlight (which is mostly visible light) passes right through them. However, greenhouse gases absorb light in the infrared part of the spectrum, the wavelengths of light that Earth's surface emits. The wavelength of infrared radiation is just a bit longer than what our eyes can detect, but like other types of light we can't see (such as radio waves or X-rays), infrared radiation carries energy away from its source. As Earth's surface is warmed by incoming sunlight, it radiates light in the infrared. Some of that infrared radiation is absorbed by greenhouse gases in the air, and the air warms up because of that captured energy. Thus, greenhouse gases form an invisible blanket, trapping heat that would otherwise escape into space.

By the time Arrhenius wrote his paper, people had measured how good CO_2 was at trapping heat, and he used these measurements to calculate how much warmer the atmosphere would be if the amount of CO_2 in the air went up (he also calculated how much it would cool if it went down). Conceptually, this is not a hard problem. All other things equal, when the concentration of greenhouse gases in the atmosphere rises (as it is doing rapidly now), the amount of sunlight hitting the surface doesn't change much, but the amount of heat that can escape goes down. This creates an imbalance, with more energy being absorbed than emitted, and the surface begins to warm. It's just like putting on a blanket while you're lying on the couch. The blanket traps some heat that would otherwise escape into the room. As a result, your body's surface starts to warm up. You will warm up until you reach a new equilibrium. The same thing happens to Earth when we add greenhouse gases to the air.

This simple understanding is a good start, but as climate detectives we need much more nuance. It's not enough to say, in theory, that doubling the amount of greenhouse gas in the air would warm the planet. We are trying to prove that the warming we are observing now is the result of our greenhouse gas emissions, which are changing rapidly, so understanding a simple doubling isn't really enough. And we know there are lots of knock-on effects that Arrhenius didn't consider. As one example among many, if it gets a bit warmer, more water will evaporate from the ocean, which makes for more clouds. Will these clouds cause cooling because they reflect incoming sunlight? Or will they cause warming because the water vapor of which they are made is itself a greenhouse gas? These so-called feedbacks mean pen and paper are not enough to get a detailed picture of what is happening to our planet. We need a much more sophisticated understanding of the climate than Arrhenius had.

Luckily, we have way more data to work with than Arrhenius did. We now monitor the atmosphere 24-7 all over the planet, using ground-, ocean-, air-, and space-based instruments. We can analyze these data because computing power has increased so rapidly. The first computerized climate models of the 1970s were on punch cards, fed one by one into room-sized computers that were far less powerful than your smartphone. Now computers can make trillions of times as many calculations as Arrhenius could with a pen and paper, so we can include many more things in our mathematical models than he could. Indeed, the climate models of today are so complex that they involve millions of lines of computer code. It takes months on a supercomputer to run a set of simulations, which solve for temperature, pressure, momentum, and chemical composition of tens of thousands of slices of the atmosphere and ocean each consecutive second. The models include mathematical representations of the reflectivity of the surface; the amount of sunlight hitting a particular part of the atmosphere at a given time of day; the difference in the heating of different parts of Earth that drives differences in temperature and pressure, and therefore the movement of air; the way that air flows over water and land, and the friction that causes; the resulting ocean movements; and many, many more things. It would take me many pages just to list what is in these models, let alone the equations that interrelate all the variables.

This makes these models sound impressive, but just because a model is complicated doesn't mean it's good. If you're going to believe a model, any model, you need to test that it adequately represents reality. Remember my claim: computer models of climate can help us do an experiment, an experiment that can test the hypothesis that human emissions of greenhouse gases are causing climate change. Should a jury really be convinced by a fancy computer program? Let's think together about why these programs are good enough to be admissible evidence in our whodunit. To do that, we need to explore what a model really is, what makes a good model, and what makes a bad one. And in our case, we are looking for a model that will help us solve our mystery and find out what is causing climate change.

———

I've made this bit its own little section because I'm going to wax philosophic about models and modeling before I come back to climate models specifically. Let's start with something everyone will recognize as a model, if not a mathematical one. Imagine a model airplane, like one you would give to a kid. At first blush it seems very different from a series of mathematical equations meant to describe and predict the climate of a planet over centuries. But if we explore a little more deeply, we'll find several important commonalities.

A model airplane represents the key parts of an airplane and omits other parts that aren't relevant for the model to be useful (to the kid). For example, the toy plane doesn't have to rely on jet fuel. No problem; the kid uses their imagination to make it fly. The toy plane doesn't need to be hundreds of feet long (impractical for the parents). But the toy plane comes with all kinds of cool stickers, so that's a plus.

Is this a good model? It depends on your point of view. The model seems to be an adequate representation of the real thing in that it's good enough for the kid to have fun with (which is the point). It's not a perfect representation of reality, but having a real airplane in the house would be worse. This model is useful even though it's "wrong." Would a model airplane with more pieces be a better model? It might look more like a real plane, but it might be much harder to build or to figure out what goes wrong when it falls apart. More complex models are not a priori better—models need to be judged by whether or not they provide a useful representation of the thing they are supposed to approximate.

The same toy model is probably not useful to an aerospace engineer working to build a better, real airplane. They need fewer cool stickers (perhaps), less of an ability to throw the plane up in the air during imaginary air battles, and more details about how the machine actually will work. They would need a really accurate representation of the interior, to see how people might need to use the space inside. They certainly would want precise representations of thrust, wing angle, lift, and so forth. To them, a model in a computer, one that represented all of those things mathematically, might be more useful. Is that a better model of the plane? For the engineer, perhaps, but not for the kid. To quote a famous philosopher of science, "All models are wrong, some models are useful" [7]. "Useful" depends on the user.

Let's move from toys to something closer to the climate models of today, which are based on physics, chemistry, and to some extent, biology. To do that, I'll start with two of the most famous scientific modelers ever: Isaac Newton and Albert Einstein. Both used math to build models of how the universe works. Newton developed a model of motion and gravity that was state of the art for hundreds of years. Newton's models were so good that we call them laws. That is as close as scientists come to calling something proven. One law, Newton's law of gravity, posits that the gravitational force between two objects is proportional to their masses. Hence the gravitational pull on a person on Earth (what we call weight) is more than what it would be if that same person were standing on the moon (because the moon is less massive). The planetarium at the American Museum of Natural History in New York City, near where I grew up, used to have big scales you could stand on to see what you would weigh on the moon. I loved them. Sadly, their replacements are less awe-inspiring (not to mention often broken). But I digress. Newton's model is brilliant in its simplicity. What's more, it's incredibly useful. It, along with Newton's other laws of motion, allowed astronomers to predict the existence of planets, like Neptune, that we hadn't yet seen. Newton's laws allow engineers to design everything from bridges and roller coasters to airplanes.

By the early twentieth century, Einstein had developed a new model of gravity, one that made a startling prediction: light (which has no mass) should also be affected by gravity. This is impossible under Newton's model—gravitational attraction requires both objects to have mass, and light has no mass. Newton is very clear on that. But a few years after Einstein made his prediction, astronomical measurements during a solar eclipse detected the gravitational effect on light [8]. Here was empirical proof. Light, which has no mass, is deflected by the gravity of the sun. Therefore, gravity does not depend on the mass of both objects. Newton was wrong. Einstein was right.

Does this mean Einstein's model was better than Newton's? Not necessarily. Einstein's model is way too complicated to use for building bridges, airplanes, or roller coasters. For that, Newton's model is all you need. Sometimes, simpler is way better.

All models are reductionist views of reality produced for a specific purpose. Einstein's is better if you're concerned with the way the universe works. For building bridges, go with Newton. All models are wrong; some models are useful. Let's turn our attention now to whether our climate models can be useful in our detective story.

———

We are looking for a way to establish that human emissions of CO_2 and other greenhouse gases are causing climate change. To do this, we need a model that is a useful representation of the climate. We can then use that model to run experiments with or without human emissions of greenhouse gases and see if the climate behaves differently under those different treatments, just as we did with the bacteria and the antibiotics, but in a computer instead of on a lab bench.

In order to build such a model, we need to include all the processes that we can describe using physics, chemistry, and math that matter for the climate. For example, we need to understand how much sunlight is hitting each point on Earth and how that changes over time (over the day, over the year, or between years). We need to know the concentration of greenhouse gases in the air and how those have changed over time. We need to know how reflective Earth's surface is, since that dictates how much sunlight is absorbed and warms the surface. A bulleted list of the things we need to know would run on seemingly endlessly. And for each thing we need to know, we would need to be able to write an equation that describes how that thing would change when all those other things change. For example, as a day warms Earth up, how much will the water warm compared with adjacent land? Will that difference drive the formation of winds and clouds, and how so? And how will that affect the amount of sunlight that hits Earth a minute later? The computer code for the most sophisticated climate models is Proustian in length.

But for all their complexity and remarkable improvement over the decades since climate change became a major concern, climate models remain constrained by three problems. We don't know everything that might be important, we don't have perfect data from every place on Earth, and computers are still too slow to solve all those equations, especially if we try to solve all those equations at very fine spatial scales. For example, most modern climate models break Earth's surface up into boxes that are about sixty miles long on each side and update the temperature of each box every second as the amount of sunlight, air temperature, air pressure, and air chemistry are calculated from the results from the previous second. Sixty miles on one side is pretty small—the first modern climate models broke Earth into boxes hundreds of miles on a side. But sixty miles is still way bigger than a cloud, a patch of broken-up sea ice, or a farm surrounded by forests and cities, all of which absorb and reflect sunlight differently. These models are wrong. How can we tell if they are useful enough to prove, beyond a reasonable doubt, that human emissions of greenhouse gases are responsible for climate change?

There are really only two ways to test whether a model is useful. The first, which is the ideal, is to use the model to make a prediction and then wait and see what happens. For example, Edmund Halley used Newton's laws to predict the return of a comet seventy-six years later. It didn't come back until after he died, but come back it did. This was a risky prediction in that the comet was very unlikely to reappear at that exact time if Newton's laws were wrong. Einstein's theories made all sorts of these risky predictions that couldn't be verified until long after the predictions were made. Although philosophers of science get tied up in knots about this, I think risky predictions that come true are very convincing tests of models. But modeling future climate change is a real problem because we want to make predictions and know whether they are right *before* those predictions come to pass. That's the whole point: to see what's coming so we can change our behavior if it's dangerous. If you are running toward a cliff, you'd like to know whether the predictions about the theory of gravity are right *before* you go off the edge.

If we can't wait for the future to tell us if we've got it right, how do we test a climate model? The rather unintuitive answer is that we test them against the past. Here's how it works. We build a computer model that incorporates all the things we think are important for understanding the climate—that long list I gave some examples from a few pages ago. We then start the model in the past—for example, in the year 1900. We tell the model the initial conditions of what Earth was like at a particular time—say January 1, 1901, at 12:00:01 A.M. We then input the amount of sunlight hitting Earth at that time, where it hits, what the atmosphere was composed of at that time, where there were forests, oceans, continents, mountains, and so forth. Given this input, the model calculates what the temperature (and pressure, momentum, and chemistry) of the atmosphere and ocean should be at every place on Earth (and in the different layers of the atmosphere and ocean) at 12:00:02 A.M., on January 1, 1901. At that time, Earth has rotated a little, so the amount of sunlight hitting each place is slightly different, and so the model solves all those equations again, figuring out what it would have been like at 12:00:03 A.M. It does this over and over, until the present (I'm writing this in 2022). Then we can compare what the model has predicted 2022 is like (or any other time between 1901 and 2022) and compare that prediction to what we observed Earth was actually like at that time.

But wait, you say. How do we know what it was really like at 12:00:01 A.M. on January 1, 1901 (the initial conditions)? We didn't have a global network of satellites observing Earth. Isn't that a problem? It turns out we can actually test that, varying what we tell the model at the start (the initial conditions), and seeing how different our predictions turn out to be. Luckily for climate modelers, climate, unlike weather, is not particularly sensitive to the initial conditions. For example, I can tell you with certainty that July will be pretty warm in New England, much warmer than January. The initial conditions of a warm or cold January have relatively little effect on whether we have a warm or cool July. Also, it doesn't matter if a particular January was unusually warm. We can be pretty sure that the following July will be warmer. Climate isn't very sensitive to initial conditions.

While all climate models rely on the laws of thermodynamics, different climate models reflect different scientists' understandings of the best ways to incorporate those laws in a representation of all of Earth. As a result, there are dozens of different models that are used by the global scientific community, each with its own set of mathematical representations of the planet. None of them are "right," in that none are perfect representations of reality. Indeed, I wouldn't bet much that any one of them was more "right" than another. But looked at together, they are very useful for our detective story.

If you start running all the models in 1901 and input all the things that have changed since then—such as burning fossil fuels, volcanic eruptions, changes in solar activity—the models do a pretty good job of predicting how Earth's climate has changed between then and now. The models get the average temperature of each region of the world right. They predict the observed changes in temperature at different layers of the atmosphere. They correctly predict the observed global cooling associated with big volcanic eruptions—for example, the 1991 Mount Pinatubo eruption in the Philippines. They simulate the observed loss of arctic sea ice as the temperature has risen. This shows us that the models do a good job of simulating the climate system [9]. This is evidence that the models are good enough to be useful.

Unfortunately, we have delayed action on climate change long enough that there is another way we can test the utility of our models. I said a few pages ago that a very good test of a model is when it makes a risky prediction that turns out to be true. I also said this wasn't a good way to test a climate model, because if the models indicate that we are dangerously altering the climate, we might want to use that information to change our behavior and thus make what happens different from what the model predicted would happen. Frustratingly, the predictions made by models in the mid-1990s can be tested this way, because we have not heeded their predictions for almost thirty years. A model from the 1990s can do a very good job of predicting what happened in the past 30 years. It's not the test I would have chosen, but sadly it's one that is very convincing. Given human emissions of greenhouse gases over the last thirty years, we are right where those models predicted we would be. The long and short of it is that climate models work well enough that we can be very confident that they will be useful for predicting the future.

Which brings us, at last, back to the scene of the crime. The climate is changing. Human emissions of greenhouse gases stand accused. We are confident our models represent reality well enough. We've tested them against real data and found they accurately re-create what happened over the twentieth and early twenty-first centuries. We can now remove human emissions of greenhouse gases and create a model world where we didn't burn all those fossil fuels and cut down all those forests. We can start all our models on January 1, 1901, at 12:00:01 A.M. and run them with natural variability (changes in sun's output and volcanic eruptions) and without anything humans have done (like releasing fossil-fuel and deforestation-derived greenhouse gases into the air). We can ask, Do these natural-only models make good predictions about what happened to the climate between 1901 and now?

The clear answer is no. Natural variation alone cannot explain the dramatic warming trends of the last fifty years. But when we include those greenhouse gas emissions, the models trace our climate trajectory extremely well. That is the smoking gun. Without human emissions of greenhouse gases, there is no credible way to explain what has happened. None of the other suspects could have done it. But when we include this one additional suspect, the human emission of greenhouse gases, we can explain what we see. Case closed.

————

I started this chapter with two goals. The first was to guide you through modeling experiments that conclusively show the human emissions of greenhouse gases are causing the climate change we are currently experiencing. Now that we have confidence that our models are useful, we can use them for the second goal: to see into the future and explore what will happen if we do, or don't, change our emissions. This works because the same physics and chemistry that determined what *did* happen will determine what *will* happen. By far the biggest uncertainty is what we will choose to do.

We can explore these alternate futures by modeling different scenarios. In one scenario, we continue to burn fossil fuels at increasing rates for decades. In response, Earth will warm something like 7°F, although we cannot rule out a much higher number. In another scenario, we eliminate fossil fuel emissions and deforestation in the next couple of decades and actively start using as-yet-untested technology to pull massive amounts of CO_2 out of the air and store it underground. In this scenario, global temperatures stabilize at something like 3 to 4°F warmer than they were during the Industrial Revolution. That's still a lot warmer, and things will be different. Coastal cities and small island nations will shrink and disappear, people will have to move, and organisms will go extinct. Limiting warming to this level will be really hard. We've emitted about 2,500 billion tons of CO_2 since 1850, most of it in the last few decades, and we now emit about 50 billion tons a year. To stay at this lower level of warming, we can only emit about another 500 billion tons; that's only ten years of emissions at current rates. Time is definitely running out.

Nonetheless, I am very sure that the lower number is worth striving for. The difference between the two scenarios is dramatic. The first, a world 7°F warmer, is a world so different from today it's hard to imagine. The second is recognizable as the one we are in today. There are many uncertainties about exactly how hot it will get and what exactly the world will look like if it does. But there is no doubt that with each degree hotter the world gets, the less and less recognizable it will become. What's more, the faster the world heats up, the harder it will be for people to live well through the transition. Our models are definitely good enough that we should heed their warning. As I said in the last chapter, because our change to the carbon cycle is almost entirely due to our demand for energy from outside our bodies, we have a chance for a less dystopian future. I'll come back to that ray of hope in the last part of the book. First, I want to return to our overarching narrative and think about the way humans are using the other elements in our story. So, for now, I will leave carbon behind and turn to the rest of Life's Formula.

5

The Goldilocks Element

TODAY, THERE ARE NEARLY eight billion people on the planet. In 1971, the year I was born, there were half that many. When my father was born in 1933, there were half again, closer to two billion. Over the course of the twentieth century, fossil fuels have provided the energy for almost all of the activities in our various and ever more numerous lives. They have powered our movement, enabled our science to improve medicine, and fueled our wars as well as our peace. Those carbon-based fossil fuels—coal, oil, and gas—were essential for the boom times of the twentieth century. But carbon and energy are not all of life's essentials, and the population explosion of the twentieth century was not solely the result of our exploitation of a new energy source. Like other world-changing organisms, our domination required more than one of life's essential elements. In this chapter, I'm going to move the focus away from the carbon-based revolution in human energy use and explain how we unlocked the power of nitrogen. This element, perhaps more than any other in Life's Formula, underpins the leap from two to eight billion people that occurred in just a couple of generations.

The story of nitrogen has both differences from and commonalities with that of carbon. First, a key difference: Fossil

fuels and the carbon-based energy they provide are finite—
there is only so much coal, oil, and gas buried for us to find.
Nitrogen is very much the opposite. It is infinitely abundant (or
virtually so) in the air. The trick is getting it. As we saw in the
chapter on cyanobacteria, nitrogen in the air is virtually inert—
two nitrogen atoms are bound so tightly together that it takes a
lot of energy to break them apart. Still, there is more nitrogen
in the atmosphere than all the organisms on Earth could ever
use, especially because there are other organisms that recycle
the nitrogen organisms use and send it back into the air.

There is another difference between carbon-based fossil energy and nitrogen. Nitrogen is completely irreplaceable—
carbon-based fossil energy is not. We primarily use carbon
outside our bodies as an energy source. But we need nitrogen
inside our bodies to make cells. Amino acids, the building blocks
of all proteins, are so named because they contain an amine
group (NH_2), some carbon, and some oxygen. Since the chemistry of life is basically the same across all living creatures, the
more organisms there are, the more nitrogen is needed. There is
no substitute. To be clear, there is no substitute for carbon in our
bodies either. It's just that most of the carbon we use is for energy outside our bodies. But the nitrogen we need, at least the
majority of it, is needed inside us and the things we eat.

The differences between carbon and nitrogen are important,
but the two elements also have some things in common. They
both have inorganic (nonliving) forms with environmental
ramifications. For carbon the two most important are carbon
dioxide and methane, both of which are greenhouses gases that
we spent a lot of time on in previous chapters. The list for nitrogen is much longer. It's a dizzying array of very reactive molecules that can readily change from one type to another (and
back). Here's a roster of the important players.

Ammonia (NH_3) is the building block of amino acids and proteins, a key plant nutrient, and also a cause of harmful air pollution. Nitrate (NO_3^-) is another key plant nutrient, but it is also a carcinogen and a major component of acid rain. Nitric oxides (NO and NO_2) are gases, reactive precursors of smog and ozone. Through their effects on air quality, these N gases thus have huge impacts on human health and play a role in over six million premature deaths worldwide. Another gas, nitrous oxide (N_2O), is a greenhouse gas three hundred times stronger than CO_2 (it's also an anesthetic known colloquially as laughing gas). There are many more, but those are the main players with regard to the environment. Don't worry if you don't remember all the names, and don't stress about all the formulae—I'm going to lump them all together and call all of them reactive nitrogen. This is to contrast them to the two tightly bound nitrogen atoms (N_2) that make up the majority of our atmosphere but are essentially inert.

In a nutshell, here's the problem with nitrogen: humans and the crops they depend on need reactive nitrogen to make the proteins that comprise their bodies. Nitrogen is abundant in the air, but it isn't reactive. In fact, for the first four billion years of Earth's history, only single-celled organisms, cyanobacteria among them, could access the unreactive nitrogen in the air. The rest of us were dependent on them to bring new reactive nitrogen into the food chain.

This chapter tells the story of how human innovation tipped the balance from a planet where reactive nitrogen was scarce to one where many places are awash in it. Since reactive nitrogen is key to plant growth and therefore food production, having a lot of it is a good thing. But because nitrogen is also a dangerous pollutant, having too much of it is a bad thing. The Goldilocks story of nitrogen has many interesting twists and turns.

I'm going to stay true to the convoluted path, telling a story that starts with my childhood babysitter, moves to the German war effort in World War I, and then to industrial agriculture in the American Midwest. It will make sense eventually.

————

When I was a little kid, my babysitter Addie Adams used to take me out looking for four-leaf clovers while my parents were at work. When we happened to find one—and it was always she who did, I never could—she insisted that I take them home and press them between the pages of a book for good luck. She was sure that they were a powerful good luck charm.

Had we paid closer attention to roots rather than numbers of leaves, we might have noticed something truly amazing about clover (and not just the four-leaf ones). Clover roots are often studded with tiny nodules, translucent white beads attached to the thin, hairlike roots. Inside those beads, far too small to see, are some of the most important players in the global nitrogen cycle and in the history of human agriculture. A class of bacteria that can fix nitrogen from the air and turn it into reactive nitrogen that they, plants, and eventually we humans can use.

Clovers are in the Fabaceae family of plants, and along with peas, beans, and beautiful wildflowers like lupines, are colloquially called legumes. Many plants in this family and some in a few other families of plants can form symbiotic relationships with bacteria that can fix nitrogen. For convenience, I'm going to lump them together as legumes, even though some plants that pair with nitrogen-fixing bacteria aren't in the legume family and have slightly different characteristics. Legumes build root nodules that are little bacteria houses and feed the bacteria carbon gained through photosynthesis (Figure 9).

FIGURE 9. The nodules on this tropical tree root are plant-built houses for nitrogen-fixing bacteria. Photo courtesy of Joy Winbourne.

The bacteria use that carbon to grow and to fix nitrogen (i.e., capture it from the air). In return for their carbon won through photosynthesis, plants get to use the nitrogen when bacterial cells die, and they can use it to build more photosynthetic machinery. It's a great arrangement for both parties.

The bacteria need these special houses because of an accident of evolution. As I briefly mentioned in the chapter on cyanobacteria, the process of nitrogen fixation depends on an enzyme, a biological machine, called nitrogenase. Nitrogenase binds with N_2 molecules from the air and is able to split them apart, recombining them with hydrogen to form reactive nitrogen that organisms can use to build amino acids and proteins. The problem is nitrogenase is what is called a promiscuous enzyme. It doesn't just attach to nitrogen. In fact, it also binds tightly to oxygen. This promiscuity is a real problem because air is 21 percent oxygen, and nitrogenase binds so tightly to oxygen that once it does it stops working for anything else. I should say that this promiscuity is a real problem in the modern world. But when nitrogenase evolved, some three billion years ago, long before the Great Oxidation Event, there was no free oxygen in the environment. Relying on a critical enzyme that was irreversibly incapacitated by oxygen wasn't a problem before the Great Oxidation Event. But because this enzyme allowed cyanobacteria to accidentally oxygenate the planet, nitrogen-fixing organisms and their partners have spent over two billion years trying to keep nitrogenase away from oxygen.

The late evolutionary biologist Stephen Jay Gould, who wrote innumerable articles about evolution for a general audience, relished this sort of Rube Goldberg–like evolutionary solution. It illustrates the point that evolution is not a march toward perfection. Rather, the tree of life is populated with organisms, genes, and biological machinery that were simply good enough at the time they evolved to be passed on to the next generation [10]. Nitrogenase is an excellent example of "good enough." It evolved in a world without free oxygen, so being irreversibly destroyed by free oxygen was not a problem. In the more than two billion years since the oxidation of the planet, no better nitrogenase has evolved [11]. Evolution produces "good enough."

This "good enough" enzyme saddles its makers with the burden of letting nitrogen in but keeping oxygen out. To compensate, cyanobacteria build elaborate structures in their cells to keep the nitrogenase separate from the oxygen that is being produced by photosynthesis (in the same cell). And plants, or at least the plants that house nitrogen-fixing bacteria, build little bacteria houses in their roots that keep the oxygen out. Not only that, but they also line the interior of the house with a molecule that binds oxygen and keeps it away from the bacteria. That's a big investment on the part of the plant—it's using some of its resources to build stuff to protect bacteria. But it's worth it, because in exchange it gets life-giving nitrogen.

Nitrogen-fixing bacteria living in soil separately from plants may have existed on land for a very long time. But the evolution of the symbiotic relationship between bacteria and land plants occurred about seventy million years ago, more than three hundred million years after plants colonized land. Today this symbiosis often supplies the majority of nitrogen to ecosystems. The bacteria get it first, then their hosts. But sooner or later when their hosts drop their leaves, die, or get eaten, that nitrogen becomes available to be taken up by other organisms, including us.

Science did not discover nitrogen fixation until the middle of the nineteenth century, but for millennia farmers have planted legumes and plowed them into the soil to increase fertility the following year. Because legumes and their bacteria bring new nitrogen into the soil, they are providing their own fertilizer. In previous chapters I asked you to think of the air as a bank account for carbon, with plants withdrawing carbon from the atmospheric bank account via photosynthesis, and respiration and fossil fuel combustion putting carbon back in. I'm going to use that same analogy here for nitrogen, but now take the perspective of the soil nitrogen bank account. Soil stores nitrogen (and other nutrients) that crops need to grow. The crop takes that nitrogen up from the soil, making a withdrawal. If the crop dies and decomposes in the field, that nitrogen is redeposited in the soil. But when that crop is harvested, some of the nitrogen in the crop, embedded in the proteins that make the crop nutritious, is withdrawn from the soil bank account. That nitrogen has to be replaced or the soil bank account will eventually run out of nitrogen. To balance the books, farmers planted legumes to pull nitrogen from the air and replenish the nitrogen that was lost to the harvest.

Today legumes are a key source of protein around the world, in part because their access to nitrogen makes them very protein rich. Soybeans, the world's most farmed legume, are the sixth-most produced crop, and other legumes such as cowpeas, lentils, and all sorts of beans are staple foods in many cultures. Indeed, humans now plant so many legumes that we can detect their effect on the total amount of reactive nitrogen circulating around the planet. But it turns out that planting legumes, while innovative and critical to human well-being, is a rather small change relative to others we have more recently wrought to the bank accounts of nitrogen. In the early twentieth century, humans finally solved the puzzle that had eluded all multicellular organisms for four billion years—we figured out how to fix nitrogen for ourselves.

For this part of the story, we will leave behind Addie's search for four-leaf clovers, the lucky legumes, and proceed to the next stop in our tour of the history of nitrogen: the First World War. Specifically, the story starts with the British naval blockade of Germany. The blockade cut off the German supply of nitrate, a reactive form of nitrogen preserved in certain rocks, which at the time Germany imported largely from Chile. Once in Germany's world-leading industrial facilities, nitrate could be processed into ammonium nitrate, which is highly explosive and was a critical component of German munitions. Without access to Chilean nitrate, the German war effort was in trouble.

The German army faced the same dilemma other organisms have faced for the four-billion-year history of life—in the air, nitrogen is abundant but in a chemical form that is inert and unusable. Transforming it into usable forms, either food or munitions, is a tricky process. Legumes solved it by building houses for bacteria. Other than that, there haven't been any big innovations in the process for more than three billion years. From the German point of view, the British blockade meant time was running out.

A German scientist named Fritz Haber actually solved the puzzle just before the war. His benchtop process combined hydrogen with nitrogen to create ammonia (NH_3), producing an identical result to one in the reaction facilitated by the enzyme nitrogenase. In so doing, he solved one of nature's greatest puzzles. This made waves in the chemistry world, but the innovation also became important to the German war effort. Haber paired with Carl Bosch to scale up this benchtop process and alleviate Germany's nitrate shortage. The industrialization of Haber's lab experiment, now known as the Haber-Bosch Process, was critical to the German war effort. The reactive nitrogen it produced was used for munitions in the absence of Chilean nitrate. But even at the time, it was clear this innovation had myriad applications. As the war ended, Haber's initial breakthrough won the Nobel Prize for Chemistry in 1918. It was a reasonable if problematic choice, as Haber also played a key role in developing the poison gas that inflicted horrible casualties on both sides of the German front. Haber, a German Jew who eventually fled Germany and died in 1933, was a complicated man. But he left a legacy of life as well as death [12, 13].

In the relative peace since the end of World War II, the Haber-Bosch Process has played a far more important role in the world than its original purpose of creating munitions. The synthesis of ammonia is the first step toward the industrial production of nitrogen fertilizer, which has changed the world on

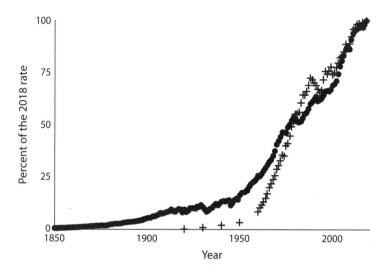

FIGURE 10. The race between two of the biggest human-caused global changes: the rise of carbon emissions (circles) and nitrogen fertilizer use (crosses) from 1850 to the present. Values have been plotted as a percent of the current rate (as of 2018). Seen in this light, post–World War II human changes in nitrogen use were even faster than those of carbon (they have a steeper slope), although carbon increases have caught up in twenty-first century. Data from the Global Carbon Project, the Food and Agriculture Organization, and Smil, *Enriching the Earth*.

a scale, and at a pace, every bit as dramatic as the human exploitation of fossil carbon. In fact, the change has been faster (Figure 10 [14]). The consequences are harder to see than disappearing coastlines, the demise of the coral reefs, and unprecedented floods and droughts that result from climate change. But they are not necessarily smaller.

When Addie and I were out looking for four-leaf clovers in the mid-1970s, there were four billion people on Earth. That was remarkable, given that for all of human history until 1800 the human population stayed well below one billion. It took only a little more than a century for the human population to reach two billion in the late 1920s. Three decades later, in 1960, it reached three billion. Fourteen years later, it hit four billion. In 2022, as I write this, the population is approaching eight billion. By 2050, it will almost certainly be close to ten. The population growth before World War II came from improvements in sanitation and medicine. Of course, both fields have continued to improve since then. But after World War II much of population growth came from our ability to feed more people. This, in turn, was a result of Haber-Bosch nitrogen.

Which brings me to part three of the modern history of nitrogen, which plays out in the industrial farms of the United States Midwest, Europe, and other breadbaskets of the world. My most recent trip to a world awash in Haber-Bosch nitrogen was to Iowa, in the midwestern United States. Iowa is a land of corn and soybean farms as far as the eye can see. Soybeans are legumes; they can do fine without nitrogen fertilizer. Corn has no such luck, and thus farmers must dump huge amounts of Haber-Bosch nitrogen fertilizer on the world's most grown cereal crop.*

Scientists estimate that half the people on Earth today would starve without nitrogen fertilizer produced by the Haber-Bosch Process. We simply couldn't produce enough food to feed everyone. As a result, the ability to fix nitrogen and increase food production may be the most important invention you have never heard of.[†] Nitrogen fixation (along with improved seed varie-

---

* Corn is the second-most produced crop in the world by weight, behind sugar cane. Most major crops are cereals, modified grasses from which we harvest and use the seed. Other examples are wheat, rice, and barley.

[†] I'm not sure who first termed it this. I learned of it first from the sustainability scientist Vaclav Smil.

ties) is at the core of the so-called Green Revolution, a post–World War II rapid increase in food production that spread like wildfire from country to country, bringing with it plunging rates of malnutrition. First in the United States and Europe, then in Mexico, then around the world, fertilizer has paved the way for previously unimaginable population growth.

The bank account analogy helps express why increased agricultural production must be accompanied by fertilizer use. As I said, every time you harvest a crop and eat it, the nitrogen (and other nutrients) in those plants, the very nutrients that make that crop good food, are taken out of the soil and moved to wherever you are. Some of those nutrients accumulate in your body (if you are growing), but most pass through. Either way, unless you and your waste are returned to the farm, there is a net loss of nutrients from the soil—a net withdrawal from the nutrient bank. When there were relatively few people, most of whom lived, went to the bathroom, and died on or near the farm, leaving fields fallow or planting legumes and plowing them under was a reasonably sustainable way to produce food. But for eight billion people? Or for ten? The food required to feed all of us requires a lot of nitrogen to be removed from farms, and it needs to be replaced, or the soil bank account of nitrogen will run out. This means feeding the world requires industrially produced nitrogen fertilizer, at least for the foreseeable future, until we figure out a safe and effective way to return the nutrients passing through humans to the farm soil whence they came. In other words, we thrive because of our innovations in capturing nitrogen. Changing our arrangement will be no simple feat.

I want to take this argument a bit further, since I think there are many environmentalists who would argue that a reliance on Haber-Bosch nitrogen is a bad thing. The process of making nitrogen fertilizer is reliant on fossil fuels. It requires a high temperature (800°F), and as of now that high temperature is achieved by burning fossil fuels. Furthermore, it combines nitrogen from the air and hydrogen from another source to make ammonia (NH_3). Currently the source for hydrogen in the Haber-Bosch Process is natural gas, which is mostly methane (CH_4), a greenhouse gas and a finite fossil fuel. While less than one percent of global natural gas consumption is used for fertilizer production, making fertilizer from a finite fossil fuel is inherently unsustainable. Nevertheless, there are so many people and we require so much food that the need for Haber-Bosch nitrogen will persist into the foreseeable future. Industrial fertilizer is here to stay, and I don't think that's a bad thing. Because of fertilizer, the percentage of "undernourishment," the World Health Organization catch-all term for various categories of not having enough food, has fallen by half in poor countries since 1970. In fact, the places on this planet where hunger is still endemic (for example, sub-Saharan Africa) are the places where there is little access to nitrogen fertilizer. And the places that have made the greatest strides toward alleviating hunger in the past half-century are those that have rapidly increased fertilizer use (Southeast Asia, for example).

Just because Haber-Bosch nitrogen is a boon to the world doesn't mean there isn't room for improvement. This high-temperature, fossil fuel–based process cannot go on forever. With cheap and abundant renewable energy, we should be able to figure out a better way to make the nitrogen fertilizer we need [15]. And the high global warming footprint of creating Haber-Bosch nitrogen is just the first of two major problems with the way humans have taken over the planet's nitrogen cycle. The second is far less tractable.

The other big problem with Haber-Bosch nitrogen, at least as it is used in most modern fertilizers, is that it is very hard to keep it where you want it. Unless it is put on soil at the exact time and in the exact amount that plants require it, nitrogen is apt to get washed away in rainstorms or used by soil bacteria and transformed into gas. It's difficult to know exactly how much nitrogen crops will need; that depends on rainfall, temperature, sunlight, and a host of other factors that farmers are guessing at. As a result, if they can afford it, farmers tend to put extra nitrogen on their fields, as insurance. If the climate conditions are just right and the crop grows like gangbusters, they don't want to underproduce because they didn't put on enough fertilizer. This makes sense from the farmer's perspective. Haber-Bosch nitrogen is relatively cheap in many (though not all) countries, and thus adding a little more than the minimum does not incur a high risk. But it also means that in many years there is extra nitrogen applied to the soil, waiting to be lost to the environment [16].

Since the end of World War II, humans have more than doubled the amount of nitrogen in circulation on land, mostly by dumping fertilizer on the approximately 15 percent of Earth's ice-free land we use for farming. The losses from this fertilizer application are truly staggering. In the United States, for every hundred units of nitrogen fertilizer created, only thirteen or so end up in our mouths as food (in the amino acids and proteins we need to survive). That 13 percent is for vegetarians. Meat production is even less efficient; only 4 percent of fertilizer ends up in our diet when we eat meat. The rest goes into the environment directly, without even passing through our bodies. And of course, even that which goes into our bodies ends up in the environment eventually.

The loss of nitrogen to the environment has been described as a "leak in the pipe," [17] and just like a leaky pipe, it can do a lot of damage that one may not immediately notice. Depending on the chemical form, nitrogen gases can cause global warming, acid rain, and smog. When dissolved in water, nitrogen in the form of nitrate is a carcinogen, fuels harmful algae growth in rivers and estuaries, and ultimately creates huge zones of oxygen-free shallow oceans (such as a large swath of the Gulf of Mexico around the Mississippi Delta). Because nitrogen is so chemically mutable, it bounces from one form to another and can have cascading negative effects. Soil bacteria quickly convert most excess fertilizer to nitrate, which is highly mobile. If enough nitrate leaches from a farm field into drinking water, it makes the water undrinkable—wells in the Central Valley of California and the American Midwest, agricultural breadbaskets, are often well above the safe limit. That same nitrate may eventually reach a river or coastline, where it can fuel algae growth. When the algae decompose, the decomposition consumes oxygen in the water, causing an anoxic "dead zone" for fish and other animals. Those anoxic conditions promote the creation of nitrous oxide (N_2O), which is a potent greenhouse gas that can stay in the atmosphere for hundreds of years. Eventually, that nitrous oxide may travel up into the stratosphere, where it destroys the ozone that protects Earth from UV radiation. It doesn't stay there forever; eventually the nitrogen may dissolve back into rainwater, fall to Earth as nitric acid (HNO_3), and start the cycle again. Thus, an atom of reactive nitrogen can continue to bounce around the world, wreaking havoc, until, eventually, a bacterium manages to turn it back into N_2 gas, at which point it returns, inert, to the atmosphere.

Nobody wants these adverse environmental effects—we want cheap and abundant food, not algae-choked waterways, carcinogenic drinking water, and bad air quality. But like other world-changing organisms before us, we have not figured out a way to enjoy the benefits of our innovations without paying the costs of their unwanted byproducts. If we used all the best technology and science at our disposal, we could probably manage to reduce nitrogen losses by a quarter or maybe a third. We'll have to do better in the future, but the science isn't there yet. For better and worse, we dominate the flows of nitrogen, the most mercurial element in Life's Formula, as much as we do carbon. We reap the benefits and pay the consequences, just as the cyanobacteria did before us.

In the last part of the book, I will come back to those consequences and the ways people are thinking about minimizing them. Before I get there though, I want to draw another analogy. Like the cyanobacteria, humans gained unprecedented access to two of the atoms in Life's Formula, carbon and nitrogen. But we've done our world-changing microbial predecessors one better—we've managed innovations in gathering phosphorus and water as well. So, for the next part of our story, we will look for parallels between us and those that did something similar— the land plants.

6

White Gold, Finite
and Irreplaceable

MY PHD THESIS was about how forests are affected by the topography of the landscape in which they are situated. As is typical of PhD theses, at least many of them, it had a grandiose and rather opaque title: "Nutrient Availability in Tropical Landscapes: Exploring Old Paradigms at New Scales." My wife Beth still pokes fun at me by asking if I have "explored any old paradigms" when I get back from fieldwork. Pretentious title aside, the work took me deep into the weeds, and I struggled to explain to my parents and anyone outside my subdiscipline why in the world they should care about it (Figure 11). Since erosion on slopes played a big role in the science I was doing, I ended up explaining that my thesis was about "how shit rolls downhill." At the time, I was only sort of joking.

With the benefit of hindsight and perhaps from lessons learned over the years since, I have a better sense of the big picture. What I was really working on, indeed what I am still working on, is how phosphorus, the least abundant element in Life's Formula, and one that does not directly control Earth's temperature, comes to be so important for how the world

FIGURE 11. The strange things we do for science. In order to sample leaves from trees growing on slopes too steep to safely descend without a rope, and being somewhat afraid of heights, I used a telescoping pole with a pair of clippers attached, duct-taped a fishing net underneath the clippers, and went "fishing" for leaves. Necessity may be the mother of invention, but it was also just fun. I'm standing on solid ground off to the right of this photo holding the "leaf fishing pole." Photo by the author.

works. I've learned that the story of how humans get phosphorus is every bit as complicated and interesting as the story of how trees do.

Modern human society has been built on several very different pillars. Each pillar has its own strengths and weaknesses, and we rely on all of them simultaneously. Carbon-based fossil fuels, which provide energy, are finite but replaceable. Nitrogen, which provides food, is infinite but irreplaceable. Which brings us to the third: phosphorus. Phosphorus has the unfortunate characteristic of being both finite and irreplaceable. That makes it a precarious pillar to depend on, but depend on it we do, because it's written into Life's Formula.

As with carbon and nitrogen, all living creatures need phosphorus to build organic molecules. Phosphorus forms part of the backbone of DNA. Cells power themselves by converting ATP (adenosine triphosphate) to ADP (adenosine diphosphate), and the P in ATP and ADP is phosphorus. Unlike carbon and nitrogen, phosphorus isn't all that abundant on Earth. It doesn't exist in a stable gas phase, so it's not in the air. It's in such biological demand that its concentration in ocean, lake, and river water is typically very low, drawn down by the microorganisms living in that water. And though it's more abundant in rocks than in water, it's still pretty scarce—less than a tenth of one percent of the average rock. For something that's irreplaceable for all life, phosphorus is hard to come by. Before I dive into how humans gained access to the phosphorus we needed to change the world, I want to set the stage by exploring how the land plants dealt with the same challenge.

———

Like most people, for most of my education I gave phosphorus no thought whatsoever. At least as far as I can remember, my science teachers in high school never mentioned it. My ninth-grade biology teacher probably mentioned adenosine triphosphate (ATP—the energy carrier for cells), but I must confess to not paying much attention in that class. Still, I'm sure that if she mentioned phosphorus at all, she never mentioned where it came from or how life struggled to squeeze it from a stone. My geology professors in college studied the history of rocks, not the history of life, and to them phosphorus was an element of little interest. As a result, until I started my PhD, my first association with the abbreviation ATP would definitely have been the Association of Tennis Professionals (go Federer!). It did not occur to me that an element so ubiquitous in living tissue might be in short supply relative to biological demand and thus play a fundamental role in structuring life on Earth.

In my first summer as a PhD student, however, I had the privilege of living in Hawai'i, which is where my advisor Peter and his wife Pamela (Pam) Matson, two of the greatest scientists of their generation, did a lot of their work. It was during this summer, in between amazing farmers markets; beaches of green, red, black, and white; and hikes to verdant cliffs spilling down into the ocean, that the importance of phosphorus was suddenly made tangible.

There were four first-year PhD students in the group that summer, and we gathered in Hawai'i to figure out what we were going to work on for our dissertations. Most of the time we were on our own, studying the tea leaves of cryptic comments by our advisors and trying to figure out what we were supposed to do. But now and again Peter and Pam shepherded us around the islands to see the sites we had read about in their papers and to learn from them how to think like scientists. These were truly formative experiences, none more so than the series of excursions we took to learn how Hawai'i could be used as a natural laboratory for understanding the importance of Life's Formula [18].

Hawaii is one of the best places on Earth to understand the challenges organisms face in gathering up essential elements and how those challenges change over geologic time. The state takes its name from the largest island, which is now also called "The Big Island." The Big Island is the farthest to the southeast and is the youngest Hawai'ian island. It is an amalgamation of five separate volcanoes and is still growing as Kilauea Volcano spews out fresh lava. The young soils formed from this lava have an abundance of elements contained in rock (like phosphorus) but lack the things that come from the air (like nitrogen and carbon). My advisor and his students had actually fertilized these young forests, which grew more when nitrogen fertilizer was added, but not when other elements (like phosphorus) were. Those young forests are thus "nitrogen limited."

The reason the Big Island is an active volcano, and the other islands to the northeast are not, is that it sits over a column of especially hot rock rising from deep inside Earth. That rising rock becomes the lava that erupts on the surface. Over time, the Pacific tectonic plate, upon which the Hawai'ian Islands sit, moves slowly northwest, carrying volcano after volcano off the stationary hotspot. Someday the active volcanos of the Big Island will drift off the hotspot, go dormant, and begin a long, slow subsidence into the sea. This has been the fate of (from youngest to oldest) the islands of Maui, Molokai, Lanai, Oahu (where Honolulu is), and Kauai (Figure 12).

It was for this very reason that our little band of would-be scholars also traveled to Kauai, the oldest island, farthest to the northwest of all the high Hawai'ian Islands. Five million years ago, Kauai sat where the Big Island sits now. Its volcano was more than twice as tall as it is today (about as tall as the Big Island is now), and it too was actively erupting. But slowly, moving at about the pace that fingernails grow, the tectonic plate on which Kauai sits crept northwest, moving the island off the hotspot. Our flight from the Big Island to Kauai took about an hour and a half. But, geologically speaking, we traveled into the future to see what the Big Island might look like after

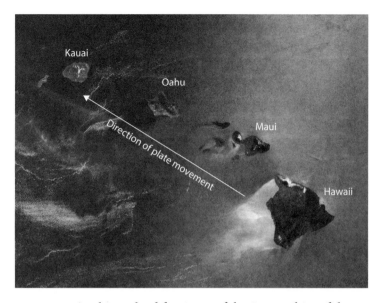

FIGURE 12. An almost cloud-free image of the time machine of the Hawaiian Islands. To the southeast, the Big Island houses the Kilauea Volcano, one of the most active in the world. Moving northwest from today's volcanic center: the island of Maui had its volcanic heyday several hundred thousand years ago; the island of Oahu, two to four million years ago; and the island of Kauai, closer to five million years ago. Each was most active when it sat where the Big Island is today, over a stationary, rising plume of hot rock sourced from deep below Earth's crust. Image courtesy of NASA.

it had sat in the warm, wet, tropical climate for five million years. In this pre–September 11 era, you could buy a ticket to this time machine at the gas station, board the plane a half hour later, and fly across five million years in the time it took to drink the little cup of guava juice they gave you on the flight.

What does the "future" hold for the young Hawai'ian forests on the Big Island? The soils on Kauai are deeper and have accumulated a lot more nitrogen than their younger counterparts.

Remember, nitrogen enters soil from the air through nitrogen fixation.* It accumulates over time. In contrast, the soils on Kauai have lost almost all of the phosphorus that was originally in the volcanic rock from which the soils were derived. Five million years of warm, wet, tropical climate has produced thick, weathered soils below which the original lava rock is deeply buried.

Plants on Kauai are stuck trying to pull phosphorus from a depleted soil with no way to get more. It's not as bad as being a photosynthetic cell in the middle of the ocean, miles from any source of phosphorus, but it's somewhat analogous. If you re-call, oceans get their phosphorus from a slow trickle of dust and river water. By tracing minerals that don't occur in Hawai'ian rocks and doing some fancy geochemical analyses, my colleagues have shown that this trickle of continental dust, blown all the way from the Gobi Desert in Asia, supplies most of the phosphorus that keeps the forests on Kauai from collapsing [18]. These ancient soils have lost so much phosphorus that phosphorus fertilizer, not nitrogen, stimulates plant growth. Thus, we say that these forests are phosphorus limited. All around the world, in places where erosion is slow enough to allow soils to weather in place, and it is warm and wet enough for weathering to occur, phosphorus emerges as the element most limiting to plant growth.

In the warm lowland rainforests of the central Amazon and the Congo, millions of years of rainfall has left deep, highly weathered soils. The geologically quiet regions have low-angle, rolling topography that limits the power of erosion to strip away the soils and re-expose the phosphorus-bearing rocks below.

---

*Recent work has shown that nitrogen sometimes is supplied from rocks, but not from Hawaiian lava, which is nitrogen-free.

The geochemical changes in the soil as it weathers creates another barrier, since the little phosphorus left in the soil is mostly bound to other soil minerals, particularly iron oxides, that have been produced during the weathering process. These rust-red iron oxides have already played a part in this story, coloring the ancient red rocks that hinted at a Great Oxidation Event billions of years ago. Those same iron oxides shape the human relationship with phosphorus as well, and that's what I want to focus on now.

If you've ever traveled in the tropics or in the southeastern United States, you've seen the red dirt that coats your shoes, cars, and everything else on a dusty day. In Kauai, tourist kiosks sell "red dirt shirts." The red sands of Tara paint the scenery in *Gone with the Wind*. All that red, the same red that painted the rocks after the Great Oxidation Event, comes from iron oxide—rust. It is formed as the iron-bearing minerals contained in the parent rock break down and reorganize in the hot, wet climate. In the oxygen-rich world we live in today, iron oxide is easy to form. It just takes time. Once formed, iron oxide is one of the most stable elements on the surface of Earth. It can last millions of years with virtually no change at all. It is important to the phosphorus part of the story because iron oxide binds so tightly to phosphorus in soil that it's hard to pry the phosphorus off.

In order to grow on ancient, iron oxide–rich soils, tropical forests have become hyper-efficient phosphorus recyclers. They have to be. If a leaf falls to the ground and the phosphorus from the leaf leaches into the underlying soil, it is likely to get stuck onto iron oxide so tightly that the plants can't get it back. In a place where phosphorus is already scarce, this is a real problem. The plants in tropical forests solve the problem by building a thick root mat, replete with millions of hair-thin roots and even thinner fungal hyphae that scavenge phosphorus from the dead leaves, recycling it before it ever hits the dirt. It's more like "reusing" than recycling, and just as in the human analog, reusing is much more efficient. In many lowland tropical rainforests, over 99 percent of the phosphorus circulating between plants and soils has been used by plants before. Almost no phosphorus leaves the system, and virtually none comes in. In the Amazon, a leaf that falls to the ground is likely to have mushrooms sprouting out of it even before it begins to visibly decompose (Figure 13). Those mushrooms are a sign that fungi are rapidly and efficiently capturing phosphorus and other nutrients from the fallen leaf before those nutrients even reach the soil.

The early part of my research career focused on forests and the phosphorus cycling within them. I knew people were changing the world—climate change, deforestation, nitrogen fertilizer—but I was pretty focused on "natural systems." That all changed a couple of weeks into my first job as an assistant professor. The year was 2007, and I was just settling into my new office, trying to figure out how to start a lab and build a career. A new graduate student, Shelby Riskin (née Hayhoe), knocked on my door and walked into my office. Shelby had already been working in the Amazon on deforestation and the conversion of forest to pastures, and she wanted my help thinking about tropical soils. I was unsure if I could be of use. I worked in tropical forests, not pastures. I knew nothing about farming and had never even been to the Amazon. But I faked confidence and told Shelby I'd be happy to help.

Then she showed me photos that took my vision of the Amazon and turned it on its head. At first, I thought the photos were of Iowa—verdant soybean fields stretching to the horizon, with a few trees in the background. Except it wasn't Iowa; it was the Brazilian state of Mato Grosso, at the southern edge of the world's largest rainforest. Shelby explained that the landscape

FIGURE 13. Mushrooms, the fruits of fungi, sprouting from a fallen leaf in the Amazon near Manaus. The fungi can scavenge nutrients, particularly phosphorus, before those nutrients ever touch the soil. Photo by the author.

in the photo had been forested for millennia, probably millions of years, before the 1980s. Then it was cleared for cattle pasture, and the forests that had survived by recycling nutrients as efficiently as anywhere on Earth were gone. It wasn't very good cattle pasture, though, because the soils were so infertile that the land could barely feed the cows. This is a typical story of deforestation in the Amazon. Once the forests are gone, farmers and ranchers can barely eke out a living because crops and cows both remove nutrients from the soil, and these soils had barely any nutrients to start with.

What was so strange about these photos, besides the giant flightless birds in the photos' foreground, was how fertile the landscape looked (Figure 14). From my thinking about intact tropical forests and the lessons of Hawai'i, I knew these soils

FIGURE 14. Shelby Riskin taking a soil core from the middle of a
soybean field that was once the southern edge of the Amazon
Rainforest. Remnant patches of forest, left around small streams,
can be seen in the distance. Photo courtesy of Shelby Riskin.

had been sitting in the geologically quiet center of South America for many tens of millions of years, losing phosphorus and
the other nutrients derived from rock to year after year of rain,
leaching, and the thickening of soils. I knew the forests were
amazing recyclers, holding on tightly to the precious phosphorus that sustained their growth and keeping it away from those
sticky iron oxides. Yet here were crops, huge amounts of crops,
grown to be exported, taking with those exports lots of phosphorus scavenged from the soil.

By the time Shelby started to work there, the southern
Amazon had emerged as one of the world's most productive
agricultural breadbaskets, burgeoning with soybeans that were
somehow thriving on the world's most nutrient-poor soils.
Shelby wanted to study this breadbasket and what it said about

the future of agriculture in the region. Her interests sparked the second big research direction of my career—one focused on the emergence of humans as world-changing organisms and the role of phosphorus in allowing that emergence to occur.

If you can beat the heat, the farm in Shelby's photos is a nice place to go for a run. The fields are divided into perfect square-kilometer blocks, so it's easy to keep track of distance on the red dirt roads, and you're likely to see tapir and jaguar tracks crossing from the adjacent forest into the fields. You never run out of room—it takes several hours to drive from one end of the farm to the other. Most farms are so big that they have to plant different varieties of soybeans at the north and south ends so they can complete the harvest before the plants rot. The farm where Shelby started working is visible on a Google Earth image that covers all of South America. It's the size of New York City (all five boroughs).

How did Brazilian farmers transform a region with some of the most infertile soils in the world into an agricultural powerhouse? The answer is fertilizer—particularly phosphorus fertilizer.

We've already discussed nitrogen fertilizer, the Haber-Bosch Process, and how the process has made humans the dominant player in the global flow of nitrogen. But soybeans, the major crop of the southern Amazon, don't need nitrogen fertilizer. They are legumes and house those incredible nitrogen-fixing bacteria in their root nodules. They get their own nitrogen from the air. The key to their success in Brazil, and indeed around the world, is phosphorus. In order to grow those soybeans, we have to give them the phosphorus they need.

Like Haber-Bosch nitrogen, today's phosphorus fertilizer is taken from nature and refined by industry. Unlike nitrogen, which is taken from the air, the raw material for phosphorus fertilizer is mined from rich phosphate ore deposits, which geological processes have concentrated in a few places on Earth. Phosphate ore, like all ore, is finite. For now, it is abundant enough, and therefore cheap enough, that modern agricultural systems have largely abandoned the conservative, recycling-oriented systems that dominate the natural world. That world has been replaced by a flow-through system.

What do I mean by a "flow-through system"? Let's follow phosphorus fertilizer to and from the Brazilian soy farms where we work as an example. High-grade phosphate ore, with more than 10 percent phosphorus bound in its white mineral lattice, is relative rare. Most of it comes from Morocco, which has over 70 percent of the world's known deposits. If you search the Web for images of "the world's longest conveyor belt," you will find an impossibly thin, sixty-mile-long line snaking its way across what seems like endless West African desert. To its south the sand is bleached white, covered with phosphorus ore blown off the conveyor that stretches from the Bou Craa mine in Morocco across Western Sahara to the ocean.* The phosphorus-rich minerals need to be dissolved to release the phosphorus before they can be applied. In the chapter on land plants, I mentioned that plants exude acids to help dissolve rocks and get at phosphorus. Humans typically use sulfuric acid in fertilizer facilities. But it's essentially the same process.

Fertilizer production requires high-grade ore because, unlike plants, humans aren't very efficient at wringing phosphorus from soil. Plants can grow on soils with as little as four thousandths of a percent phosphorus. In contrast, we start with ore that is 30 percent phosphorus. Even at 10 rather than 30 percent,

---

* Both Morocco and the Sahrawi Democratic Arab Republic claim sovereignty over the area called Western Sahara.

processing costs go up by a factor of two or three. Below that, phosphorus mining would make food production incredibly expensive.

Once produced, the fertilizer is shipped to Brazil, where it is carried by truck over twelve hours of rutted dirt roads and dumped onto farm fields in very high amounts. This occurs during the dry season, when the fields are an endless expanse of ochre-colored dirt, bounded at the horizon by the bright green, razor-straight edges of the remaining forest. When the rains come—and climate change is making that much less predictable—the soybeans shoot up. They grow rapidly for a few months, munched on by tapirs, picked over by giant flightless rheas, and sprayed with a rather terrifying array of pesticides. Most importantly for this story, they take up some of the fertilizer phosphorus the farmer added. The phosphorus-rich soybeans are then harvested by combines larger than a goodsized suburban home. The combines load the beans into eighteen-wheelers that snake their way to the on-farm drying or storage facility. During harvest, which stretches several weeks, the line can be dozens of trucks long. Once inside the drying facility, the full trucks are unloaded by a hydraulic lift that simply picks them up, flips them over, and dumps them out, in not much more time than it takes to dump out a toy truck in a sandbox—less than a minute. Soy is then stored in drying facilities the size of sports arenas—on this farm, there is one for GMO crops (to be sent to China) and one for nonGMO crops (for Europe). It's an operation of almost unimaginable scale. Indeed, one of the biggest cognitive disconnects I've ever felt was seeing five brightly colored, small recycle bins—metal (yellow), plastic (red), paper (blue), glass (green), food (brown), and a trash can (gray)—in front of a footballstadium-sized soy storage facility that was once the Amazon rainforest. "We care about the environment, so recycle your soda can on a farm the size of Paris carved out of what used to be one of the most biodiverse places on Earth."

From the farm, the dried beans and the phosphorus in them are trucked (again) back over the road whence the fertilizer came to a port on the Atlantic. From there they are shipped to Asia or Europe. In either place they are fed to animals, who are then fed to people. It is a one-way trip with no return. At the end, all that finite, irreplaceable phosphorus from Morocco has traveled around the world only to end up in sewers, rivers, and eventually the oceans off Europe and China. The whole one-way trip takes about a year. But, just like fossil fuels, what we use in a year will take tens of millions of years to re-form.

There's another part of the phosphorus story that has a fossil fuel analogy. For all the effort to get the stuff, we don't get a very good bang for our buck. Even the most efficient car engines turn only 40 percent of the stored chemical energy in gasoline into energy that can be used to turn the wheels—the rest is just lost to the environment. Phosphorus has a similar problem. Remember those iron oxides I brought up a few pages ago? The ones that make tropical soil red? The ones that are sticky for phosphorus? It turns out that when phosphorus fertilizer is applied to bright red fields of what used to be the Amazon, around half of it just sticks onto the soil minerals, bound so tightly that plants can't get at it. Our research group has looked at this in some detail. These soil minerals are so voracious that farmers have to add double the amount of phosphorus their plants need onto their fields every year. For now, phosphorus is relatively cheap, and farmers can still stay in business even as they pay this "tax" to the soil. But even at low prices, phosphorus fertilizer costs are about a quarter of the total expenses for running a farm.

What would happen if we ran out of phosphorus ore? Or if the price of getting it spiked? We've already been through some booms and busts. The earliest fertilizers came from guano, a word with Andean roots that means bird and bat excrement. People in the Andes probably used guano as fertilizer for at least a thousand years before Europeans made it to the Americas. Europeans finally realized the importance of guano-based fertilizer in the early nineteenth century, after the most famous scientist of the age, Alexander von Humboldt, traveled to Peru and wrote about its use [19, 20]. The race was on to find more guano. It was mined from remote islands covered with sea bird colonies. Sea birds soar above vast swaths of ocean, gobble up fish and algae rich in phosphorus, and concentrate phosphorus-rich excrement on their rocky homes. Built over millennia, rookeries provided the nutrient input farmers required. But like most gold rushes (guano was called "white gold"), the rush for guano decimated sea bird populations and then went bust.

Fortunately, guano is not the only concentrated form of phosphorus. In 1841, Victorian scientist John Bennet Lawes found that cabbages on his estate in Rothamstead, England, grew better when he added ammonium phosphate. He quickly patented a process for treating bones (there is a lot of phosphorus in bones) with sulfuric acid to produce "super phosphate" and started producing phosphate fertilizer in 1847. Unfortunately, bones contain phosphorus that originally comes from the soil (the plant takes up phosphorus, the animal eats the plant), so like all organic material, bones can help with recycling but don't increase the total supply available. And with more and more people living far from where they were excreting, the phosphorus cycle that predominated in natural systems was getting less and less cyclic and more and more linear.

There are sources of concentrated phosphorus ore scattered around the world. In the United States, the first ore deposits were discovered and exploited in South Carolina right after the Civil War, and by the end of the nineteenth century, major deposits were being explored in West Africa. We are still dependent on those deposits today, with little regard for the fact that they cannot last forever.

Sadly, we have largely ignored the guano lesson—that phosphorus is finite. We exploited guano until it, and the birds that left it behind, were largely gone. Then we moved on to a new source—phosphate ore, concentrated in a few places over millions of years by geologic happenstance. Commercial mining of Moroccan deposits (the world's largest) began in 1922 but didn't really ramp up until after World War II. In the postwar boom, the growing need for food led to a fourteen-fold increase in phosphate mining between 1945 and 1980. The United States and Soviet Union led the way, but their combined reserves are tiny compared with those in West Africa. That long conveyor belt across the desert is part of a massive phosphorus delivery infrastructure that keeps us all from starving, one that will increasingly depend on West Africa as other deposits dwindle.

Countries that don't have oil, gas, or coal, and even those that do, can invest in renewables and wean themselves off these finite deposits of stored energy. But there is no way to wean ourselves off phosphorus, and so the concentration of high-grade ore in just a few countries opens up some interesting questions about the future of the food supply and the power to control it. Long before the world runs out of high-grade ore deposits, the geopolitical implications of the distribution of phosphate ore will likely be something to be reckoned with. For now, these deposits are sufficient. But in the end, we will need to leave behind our linear agricultural system and return to one where this irreplaceable, finite element is recycled as quickly as it is consumed. There is no escape from the constraints placed on us by Life's Formula, but there are things we could be doing better. I'll come back to those in the last section of the book.

———

In the chapter on fossil fuels, I argued that humans were unique in our pace, if not our chemistry, of how we were changing the world. The same is true for phosphorus. Our one-way phosphorus street is unprecedented in both amount and pace. But we are not completely without precedent. Viewed in a certain light, our mining is basically a modern analog of the strategy of the land plants. Plants and fungi moved to land and began to capture sunlight that was otherwise not being used by organisms. In order to utilize that extra energy, they mined the land for nutrients and water, and used acids to break down rock and free phosphorus. Indeed, that strategy is part of why they succeeded on land four hundred million years ago, and why, on average, photosynthesis is more efficient on land than in the ocean, where nutrients like phosphorus are even scarcer. Pour acid over rock, and it dissolves, liberating life's rock-derived regulator. It's a strategy that worked for the first tropical forests over tens of millions of years before they fell victim to their own success. It has been working for humans for the past seventy years. Like so much about our success, however, there are lessons to be learned from the deep past that are reason for pause. This kind of innovation comes with enormous risks.

The biggest long-term risk with phosphorus is that we use it on human timescales, but Earth only provides it at geologic timescales. But the way we use phosphorus today presents a more immediate threat. Humans have vastly increased the amount of phosphorus moving around the planet. And as we saw in the nitrogen chapter, adding a whole bunch of a formerly scarce, critical elements to the environment always shakes things up. The same is true for phosphorus.

To show how, I'm going to take you back to the midwestern United States in the 1950s and 1960s. It was a time of massive water and air pollution. Rivers were catching on fire. And in formerly beautiful lakes dotting the upper Midwest, clear blue water was transforming into a green, slimy soup dominated by cyanobacteria and algae that seemingly sprang from nowhere. The question was: Why?

Even pond scum can't miraculously appear. If it's going to bloom, it has to gain novel access to some limiting elements. At the time, soil erosion was rampant, and one group of researchers thought that all the carbon and nitrogen stored in soils that were eroding off farms might be fueling growth. This view was supported by the detergent industry, since the other potential culprit was phosphorus, which was the main cleaning agent in detergents. The first scientific studies were small. People collected lake water in jars; added carbon to some, nitrogen to others, phosphorus to others; and saw which ones turned green with life. Their approach was very similar to the ideal experiment I described in the chapter on climate modeling. Treatment and control—nice and tidy. The results of one such study, published in *Nature* (the world's premier scientific journal), concluded that carbon, not phosphorus, was the culprit. It was funded by the Soap and Detergent Association of New York. But it did not convince everyone, as there was much to criticize about the paper beyond the source of funding.

In science, as in most of life, certainty almost always comes only from the accumulation of evidence. But in this case, it was really one experiment that solved the mystery. In the early

1970s, a scientist named David Schindler decided to tackle the question in a different way. Slimy lakes had become a big issue, and Schindler wasn't interested in small-scale experiments for a big problem. As a Canadian he had the advantage of having many lakes in his neighborhood. His team decided to fertilize whole lakes—forget the jars. They did several experiments, but the most memorable was when they cut a lake in half with an impervious barrier across a narrow constriction. They added nitrogen and carbon to one side of the barrier and nitrogen, carbon, and phosphorus to the other. The results were blindingly obvious (Figure 15). Dump a bunch of carbon and nitrogen in, and the lakes stayed beautifully clear. Dump in carbon, nitrogen, and phosphorus, and they turned the color of pea soup. It makes sense—cyanobacteria can get their own carbon and nitrogen. But phosphorus is hard to come by.

In much of the world, Schindler's experiments led to regulation that took phosphorus out of detergents, and things began to improve. But today, fresh waters the world over are once again turning green, though now it's more from fertilizer leaching off farms and animal-feeding operations. In some places phosphorus is the key culprit; in others nitrogen may be equally or more important. It depends on the water, the currents, and the details. But the overriding picture is clear. Photosynthetic organisms in lakes, rivers, and even the ocean have plenty of sunlight and water. Give them the rest of the elements they need, and they will bloom.

Unfortunately, algae and bacterial blooms are not just a problem because they make the water look bad. Some cyanobacteria produce toxins that make it dangerous to swim or to eat anything that lived in a bloom. And beyond that, the microscopic critters whose proliferation poisons the water are as dangerous in death as they are in life. When a pulse of nutrients promotes

Carbon + Nitrogen

Carbon + Nitrogen + Phosphorus

FIGURE 15. Even in black and white, the algae bloom that emerges when phosphorus is added to a lake is apparent in the bottom half of this photo. Like the bottom, the top half of the lake received carbon and nitrogen, just no phosphorus, and its water remained crystal clear (black in this photo). An artificial barrier, built for the experiment, separates the water on the two sides. Photo by IISD Experimental Lakes Area and used with permission.

an explosion of life in the water, the proliferation of life is followed by a proliferation of death when the nutrient supply diminishes. When the microscopic creatures that sucked up those nutrients eventually die, they sink to the bottom and start to decompose in the reverse reaction to photosynthesis. To remind you, photosynthesis produces oxygen. The reverse reaction, respiration, consumes oxygen. When those critters die and decompose at the bottom of the body of water, if the water conditions are right, the process can suck up all the oxygen. This means no animal, fish, shellfish, worms, or other multicellular organisms can live there until the currents bring enough oxygen-rich water to bring the system back to life. These regions created by too much life and followed by too much death are called "dead zones."

Some of these dead zones are small, but some are enormous. The dead zone in the Gulf of Mexico, fueled by excess nitrogen and phosphorus running off farms and into rivers, can be as large as the state of New Jersey (approximately 8,000 square miles). There are now episodic or permanent dead zones downstream from virtually every large-scale agricultural region in the world. They have huge impacts on marine life and on the people who make a living fishing. Dead zones are an ever-growing symbol of our access to the elements in Life's Formula.

Today, such dead zones are restricted to shallow waters like the nearshore Gulf of Mexico. But some scientists think that the amount of phosphorus flowing off our farms, through us, and into the world's oceans may trigger an ocean-wide dead zone. This would be so catastrophic that it is hard to conceptualize, although it appears to have happened occasionally in the geologic past. In the chapter on cyanobacteria, I said that the availability of nutrients that come from rock weathering (like phosphorus and iron) are probably what limits the majority of photosynthesis in the ocean. That's because there is only enough light available in surface waters, and thus photosynthesis across much of the open ocean depends on a slow trickle of rock-derived elements from dust and rivers.

So, what happens when the amount of phosphorus and other nutrients eroding off the continents suddenly jumps up? In the case of phosphorus, the rate of delivery to the ocean is probably double to triple what it was a few hundred years ago. Is this enough to change the ocean in some profound way? We simply don't know how close to the edge we are teetering. But, as I said, world-changing innovations come with substantial consequences. We shouldn't presume we are smart enough to anticipate all of them.

Of late, climate change scientists have been talking a lot about what they call planetary boundaries and alternative stable states [21]. Here's a metaphor to explain the ideas. Think of Earth as a ball sitting on a big flat table. You can push that ball around a lot, anywhere on the table, and nothing much happens. But push it over the edge, and the ball falls to a new stable state. It will never return to the tabletop on its own. We're not sure that adding all of this phosphorus to the world's oceans will push us over the edge into a new anoxic state. But we do know that we're pushing the system around a lot, faster and harder than it has been pushed before. Since we don't know where the edge is exactly, I would argue that we should err on the side of caution and push less. In the case of phosphorus, that means recycling a lot more and dumping a lot less into the ocean as waste.

Climate change and our alteration of the carbon cycle can largely be slowed, and eventually stopped, by political will. Whether we do it or not remains to be seen. It certainly has taken too long already. Some, but importantly not all, changes are unstoppable at this point. The problem of nitrogen is twofold—too little in some places and too much in others. But at least nitrogen is inexhaustible, and microbes will eventually return excess fixed nitrogen to the atmosphere. Phosphorus is different. It is finite and irreplaceable. Yet we continue to throw it away. Once again, the byproducts of our chemical success will have consequences. This is always the case. But before I turn to potential solutions, we have two more elements to explore, and they come in a package. Let's turn to the story of water.

7

Water, the Key to Life on Land

WHEN I WAS in my mid-twenties, I went back to school to get an MS degree in geology from the University of Montana (after having been a history major as an undergraduate). A year of ski-bumming in Wyoming had convinced me the Northern Rockies were one of my favorite landscapes in the world, and I jumped at the opportunity to do outdoor science in a place that had become so dear to me. As I finished up my degree, my wife Beth was accepted to medical school in New York City. I was a bit heartbroken to leave the mountains and go back to New York, so for consolation we decided to take a six-week drive around the western United States before returning to the urban canyons where we both grew up.

I had explored much of the Northern Rockies, but the deserts of the Southwest were new to me. Endless arroyos hinted at water but disappointingly revealed themselves as bone-dry, cobble-filled canyons as they zipped by the car window. For two East Coasters recently transplanted to the fairly wet Northern Rockies, the aridity was unsettling. We were avid hikers, but here it felt like a walk into the wilderness would be perilous. Not because of bears, cold, or steep topography, but simply because of thirst. It was a new feeling. If you're used to having water around, its absence is profoundly unsettling. In a way, that feeling is embedded in the start of Life's Formula.

Thus, for the final chapter of this section, I'm going to focus on hydrogen and oxygen in the form that matters most for all life—water. But the real story of water and humans isn't about the worry of not having enough to drink. That is, at the risk of sounding coy, just a drop in the bucket. It's a story about plants, their physiology, and how that physiology dictates that we "eat" way more water than we drink.

———

To show what I mean when I say we eat water, I want to start with the land plants. After all, they are the base of the food chain upon which we and pretty much all living creatures on land depend. The crops we eat most are modified grasses (corn, wheat, rice, and barley), legumes (soybeans), or starchy roots (potatoes). All have been genetically altered by selective breeding for thousands of years to improve their taste, productivity, and resilience. Many have also been modified more recently using modern genetic techniques. But despite all these modifications, our crops depend on the same basic photosynthetic processes as their wild land plant ancestors, who, in turn, inherited it from the cyanobacteria and their relatives. All use energy from sunlight to combine carbon dioxide (CO_2) and water, and this reaction provides the carbon-based molecules we all eat.

In previous chapters I talked a lot about how nitrogen and phosphorus can limit the amount of photosynthesis that happens in a given place, but on land the first-order constraint is water. If I ask you to imagine how many plants grow in the desert, you don't need to ask me how fertile the soil is in order to know that the answer is "fewer than in wetter places." As I've described earlier, it took tens of millions of years for land plants to evolve the capacity to move out of river valleys and swamps where water was abundant. Now, humans are coaxing the descendants of those plants to proliferate across the globe. We're growing crops from the deserts of Saudi Arabia to the rainforests of Brazil to the often-frozen plains of Saskatchewan. Where it rains enough, we rely on the rain. Where it doesn't, we irrigate. There is no food without water.

I want to explore the link between food and water a bit more carefully, since it's a great example of how Life's Formula places profound constraints on what is and isn't possible. Our crops grow via the chemical reaction we focused on first—the one for photosynthesis. I reproduce a slightly modified version here, with a different emphasis:

carbon dioxide + **water** + sunlight $\rightarrow$ plant tissues + oxygen

For the microscopic photosynthesizers of the ocean, which are bathed in water in which CO_2 is dissolved, it's no problem to get water. But land plants have a harder problem to solve. Water is in the soil, and photosynthesis happens in the leaves. Plants have to pull water out of the soil, all the way from their roots to their leaves. How do they do that?

The way it happens is actually kind of amazing, and the key to the puzzle can be seen (with a microscope) on the underside of every leaf. Leaves are punctuated by tiny pores, called stomata, which serve as conduits for CO_2 to diffuse from the air into the leaf. Why does this happen? When the plant is photosynthesizing, it is drawing the concentration of CO_2 in the leaf down—turning that CO_2 into plant tissue—thus the concentration of CO_2 in the leaf is lower than that in the outside air. As a result, when the pores are open, CO_2 will diffuse from the air into the leaf. It's a law of nature that gases diffuse from places with a higher concentration to places with a lower concentration. The bigger the difference, the more the diffusion.

What about getting water up to the leaves? It took scientists a while to figure this out, but we now basically conceptualize of a tree as a long straw. One end of the straw is the root, growing down into the soil where it is (relatively) wet. The other end is the leaf, dangling in the air where it is (relatively) dry. As water evaporates out of the leaf (the scientific term is "transpires," to differentiate it from evaporation that happens without going through plants), it creates a pressure gradient between the wet root and the drying leaf. Since things move from high pressure to low pressure environments, the water goes up the plant. It's similar to the process that allows you to pull water (or, if you prefer, a milkshake) up a straw. You're creating a pressure gradient, with lower pressure at the top of the straw than at the bottom. It's incredible to think of a redwood tree, almost 400 feet tall, using this simple method to pull water from the soil all the way to its leaves at the top of the tree. But that's what happens.

To recap: in order to get CO_2 into a leaf, plants open those tiny pores in their leaves, and CO_2 diffuses in while water diffuses out. For every molecule of CO_2 plants succeed in capturing through their stomata, they lose thousands of molecules of water out the same holes. Furthermore, the stomata are operated by special "guard" cells that open and close when they fill and empty of water. If the leaf has enough water, the guard cells fill, puckering like lips, and the pores open. If the leaf dries, the cells collapse, and the pores close. Much less water escapes from the leaf when the pores close, but CO_2 can't diffuse in either, so the plant will starve for carbon if they leave the pores closed too long.

In this way, water is linked to photosynthesis on land in two fundamental ways. First, evolution has dictated that land plants use water in the photosynthetic reaction itself. No water, no chemical reaction. But second, and much more importantly in terms of how much water they use, plants lose water whenever they open their pores to allow CO_2 to diffuse in. They need CO_2—they can't grow without it. But they can't get CO_2 without losing water out the same pores that allow CO_2 in. And if they lose too much water, their pores close up, and they can't get more CO_2. No wonder there are more plants where it is wetter.

Since their emergence onto land, plants have devised all sorts of evolutionary innovations to modify (somewhat) the connection between CO_2 intake and water loss. Some grasses, including some that are very important crops such as corn, use a slightly different photosynthetic pathway to concentrate CO_2 in their leaves and reduce water loss. Cacti and other succulent plants have gone to further extremes and have evolved the ability to take in CO_2 at night, when it's cooler and they lose less water, and then run the photosynthesis reaction during the day, when it's hotter but sunlight is available. Regardless of these innovations, if the soil gets too dry for too long, plants can't survive. Water and photosynthesis are inexorably linked. With this in mind, we can understand why farming accounts for almost 80 percent of human water use. It is the price imposed by the evolutionary history of our crops.

———

As a result of this evolutionary connection between water and photosynthesis, human societies stayed close to water for most of their history (as land plants did when they first emerged). But like land plants, we have now innovated, if not exactly evolved, intricate plumbing mechanisms to keep water flowing even in very dry places far from water sources. We grow water-intensive crops such as lettuce and cotton in the deserts of Arizona, fed by canals draining water from faraway rivers. Bright green crop circles stand out in stark contrast to the taupe-colored, unvegetated soil in satellite images of North Africa and the Middle East. These farms are fueled by ancient groundwater pumped up from hundreds of feet below the surface. We have moved mountains, flooded valleys, and replumbed continents to quench our thirst and that of our crops. We have produced extraordinary feats of engineering—giant dams, endless wells, and a network of canals uncounted in combined number or length. The longest, in India, is four hundred miles long, even longer than the phosphorus conveyor belt in Morocco and the Western Sahara. In our rapidly urbanizing world, it's easy to lose track of just how much land we use for food production and how much water all this production requires. But if you put all the world's farms next to each other, they would cover an area equivalent to South America. All those farms are growing plants, and all those plants need water to grow.

I find it hard to come up with meaningful analogies for how much water humans use. It is about four trillion cubic meters of water per year—an amount I can quantify but can't really conceptualize. That volume could cover the entire United States (including Alaska and Hawaii) in a layer of water almost half a meter (over a foot) deep. If you prefer to imagine height rather than width, that much water could cover Delaware and Rhode Island with water higher than the Empire State Building. Whatever your measuring stick—it's a lot of water.

Where do we get all that water? There are two main sources, surface water and groundwater, though a third, desalinated ocean water, is likely to become increasingly important in the coming decades. Surface water is just what it sounds like—water in lakes and rivers on the land surface. This water is only a small fraction of fresh water; there is much more stored in groundwater, trapped between grains of soil and rock. But in most places surface water is easier to get. Almost every river in the world of any size is being diverted, used in place, or stored behind dams. Of the biggest rivers, only the Amazon, the world's largest in terms of the amount of water it transports, remains undammed. Its tributaries—the Madeira, the Tapajos, and the Xingu, themselves more voluminous than most other rivers on Earth—have recently seen the first of many proposed dams constructed. Modern dams are enormous and change the rivers above and below them for hundreds of miles. The world's largest, the Three Gorges Dam in China, is almost a mile and a half across and stands fifty stories tall.

In wet regions like the Brazilian Amazon, dams are built to generate electricity and make waterways more navigable for barges. But in drier regions, like the American West, dams also provide water for irrigation. In such places, so much water is diverted to farms that the rivers peter out long before they reach the ocean. The Colorado River, which carved the Grand Canyon and thus literally shaped the American West, is one of those. There are so many dams and reservoirs on its course that about a sixth of its original flow simply evaporates off the reservoirs stored behind the dams. The rest is parceled out, meticulously and litigiously, to a growing population and increasingly water-hungry farms. In some parts of the Colorado watershed, it is illegal to collect the rain that falls on your own roof, because that water isn't yours. It's already been allocated. The once-mighty river now only occasionally reaches its natural outlet in the Gulf of California, although in 2019 restoration efforts allowed it to do so again for the first time in sixteen years.

So much water is captured from rivers and distributed onto farm fields that humans have effectively made the continents wetter. It's our way of changing the amount of rainfall a place receives. Rather than water falling from the clouds, it falls from irrigation pipes onto our thirsty crops. If you take all the water extracted for irrigation, and you compare that to the amount of rain that naturally hits the global land surface, you'll see that we've increased the amount of water hitting the land by a few percent. That may not seem like a big deal, but remember that that is compared to all the rainfall that falls everywhere, including over all the world's forests and grasslands. And we've decreased the amount of water returning to the ocean in rivers by a similar amount. We make a measurable dent even in those huge volumes.

Zoom in a little, to the swaths of the land that we farm in dry places, and you'll see that our influence is much more pronounced. Let's take one of the world's breadbaskets, the Central Valley of California, as an example. It's roughly four hundred miles long and fifty miles wide, bounded by the Sierra Nevada Mountains to the east, the Cascade Mountains to the north, and the California Coast Range to the west. The Central Valley grows a quarter of the United States's food. I used to drive across it frequently when I was a grad student seeking to escape the crowds of Silicon Valley and take refuge in the snowy Sierra Nevada. I drove through two hours of cows and crops before hitting the foothills—the smell of fertilizer and manure in the air the whole way. The valley is wetter in the north and drier in the south, but as a rough average the region gets about a foot of rainfall a year. But dams in the Sierra Nevada and diversion from as far away as the Colorado River add roughly another two feet. Irrigation effectively triples the amount of water falling on this enormous valley. While the Central Valley is an extreme case, the story is not so different in many agricultural regions around the world. There are no more important atoms for life than H and O. How could we not be world-changers when we are altering the availability of those life-giving elements by so much over such big regions?

Rivers are but one of the two major sources of water that we use. Though they shape everything from the location of cities to global commerce to our artistic imagination, there is actually much more fresh water stored underground than there is running over the surface. Soil and even rock are not as solid as they seem. Tiny spaces between grains can store water, creating zones from which water can be withdrawn. In some places, topography and hydrology dictate that groundwater emerges at the surface, creating springs that bubble up from the ground. In others, the groundwater remains deep below the surface and needs to be pumped up if we are going to use it.

Using groundwater has many advantages. It is harder to pollute than surface water, since soil can act as a filter and remove both disease-bearing organisms and human-made contaminants. In many parts of the world, groundwater is the only safe water to drink without treatment, and where treatment is prohibitively expensive it is therefore the only safe water to drink. Today, groundwater provides water to about two billion people and supplies about 40 percent of the total water used for farming. In arid regions this amount can be much higher, approaching 100 percent in really dry places.

This source of H and O has been just as dramatically altered by humans as the world's rivers have. Once again, the analogy of a bank account is useful. I used the analogy for our taking carbon from the slow carbon cycle and shunting it to the fast carbon cycle. I used it to describe nitrogen, phosphorus, and soil fertility. This time I want to apply it to groundwater. This time the income, the source of groundwater, is rain and snow. When rain falls and snow melts, some water runs over the surface to form rivers. But some percolates through the soil, into the interstices between grains of sediment and rock, and is stored—as if in a sponge. This stored water is groundwater, and the sponge is called an aquifer. Just as in a bank account, the amount stored depends on how much snow and rain percolate in, how much storage capacity there is in the sponge, and how fast the water leaves. In natural conditions, groundwater flows underground in much the same way it does overland, albeit much more slowly, and eventually spills out into rivers or into the ocean, where it is lost from the groundwater bank account. If rain and snow are the income, that flow out is the withdrawal.

There are vast bank accounts of groundwater. But in the driest places, where we rely on them most, most accounts had very small incomes or withdrawals before humans entered the picture. These places represent a vast water trust fund, accounts filled by a slow trickle of savings over millennia or by a vast inheritance from the last ice age, created by an income that petered out ten thousand years ago. For example, there are gigantic aquifers under the Sahara Desert that filled five to ten thousand years ago when the Sahara was a verdant grassland dotted with lakes. Others, like the giant Ogallala Aquifer under the high Great Plains in the United States, are refilled by rain and snow that still fall on the surface above them.

Even as the water income to groundwater accounts has shrunk in many places, withdrawals have skyrocketed because humans figured out that our vast inheritance could be spent. Not surprisingly, our accounts are emptying. Take the example of the biggest aquifer in the United States—the Ogallala. It spreads from South Dakota, where it is more than one thousand feet deep, to north Texas, where it is less than one hundred feet deep, and thus underlies almost two hundred thousand square miles of farmland.

After the dust bowls of the 1930s, farmers in the region were understandably desperate to maintain the water supply to their farms. Rainfall is always uncertain, all the more so in regions that are on the edge of being too dry to farm, like the prairie above the Ogallala. After the Second World War, two new technologies became available that made the Ogallala a lifeline for the farmers of the High Plains—gas engines to pump up water from below ground and center pivot irrigation to spread it around. In a sense, groundwater irrigation is our version of the evolution of roots and fungal symbionts in plants. It allows us to grow (food) in places far from the sources of surface water by extracting water held in the soil and rock beneath us. That's basically the same innovation land plants made almost four hundred million years ago, and it allowed them to proliferate across the continents.

Drop a well down to an aquifer, pump water up, attach a rotating arm to the pump, and voilà—circular patches of wet, fertile soil as perfectly round and green as a giant monochromatic game of Twister. If you fly from east to west over the United States today, you'll see these crop circles appear in great numbers shortly after you pass Wichita, Kansas (you can also "fly over" them using Google Earth). To the east, there is enough rain that farmers have less of a need for irrigation. A summertime view from above is mostly green. On the ground, the farm fields stretch as far as the eye can see, but they are rectangular, bordered simply by property lines and ruler-straight roads on the flat landscape. Continue west past Wichita toward the Colorado border, into the rain shadow of the Rockies, and the view from above turns a brownish hue, even in summer. The fields are green, but they are increasingly dominated by circles, not rectangles. These circles are the manifestation of center pivot irrigation, fed by water sourced from deep below the surface. About 80 percent of the water used in Kansas, almost all of it for pivot irrigation, is drawn from the Ogallala. It's just too dry to farm productively using rainwater alone.

The Ogallala provides water to about a quarter of all the farmland in the United States. Inevitably, as withdrawals have increased and income has not, the account is shrinking, and the water level is dropping deeper and deeper below the surface. In some places the top of the Ogallala is over a hundred feet deeper than it was when irrigation began in earnest in the 1940s. With each year, as the water drops, it gets harder and more expensive to drill down to the water and pump it up. So far, we've used up about 10 percent of our inheritance stored in this giant aquifer. That number grows every day. Even though water is a "renewable" resource, you can't make withdrawals faster than you make deposits. We are mining the groundwater of the Ogallala as if it were oil. Oil is replaceable—we don't want oil, we want energy. Water is not replaceable. Those tiny pores in leaves, the ones that let CO_2 in as water escapes, means that without water we can't grow food.

———

The Ogallala is one example of the human mining of groundwater, and thanks to a remarkable new technology we have recently learned just how representative this example is. High in orbit above Earth are twin satellites that together are called GRACE (Gravity Recovery and Climate Experiment). These thousand-pound, trapezoidal twins circle Earth rapidly, measuring where they are relative to each other and the ground. When one of the twins passes over a full aquifer, it is pulled slightly toward Earth by the mass of the underground water, and its trailing twin can detect the drop. If the satellite dips less on a later pass, the amount by which it dips less can be related to the amount of water that has been withdrawn from the aquifer below (the amount of rock doesn't change from year to year). GRACE was launched in 2002 (and was replaced by a newer version in 2017), and it maps the entire world every month. Thanks to this extraordinary mission, we have monitored the world's biggest aquifers (including the Ogallala) for almost two decades. Several recent papers have used data from GRACE to show that the majority of the world's biggest aquifers are losing water to the relentless thirst of modern agriculture.

This brings us almost up to date on the story of humans and water. People have used our engineering acumen to make sure water is available where we need it by storing water behind dams, shunting it away from rivers in canals and irrigation ditches, and pumping it up from below the surface. All of these changes allowed us to plant crops which capture the sun's energy and turn it into food, a process that cannot proceed in the absence of water because of the fundamental way in which leaves are constructed. We have engineered access to water, but there is no way to engineer around the constraints of Life's Formula.

Before we move to discussing the future of water in the era of climate change, there is one more issue that bears discussion— and it comes back to the problem of needing fresh, non-salty water. Again, though, I want to talk about this in the context of the plants we grow for food rather than our need for fresh, non-salty water to drink. At the root of the issue (pun sort of intended) is that all water, even fresh water, has tiny amounts of salts (like sodium chloride) dissolved in it. And when we use that water to irrigate our crops, that becomes a problem.

To think about why, I want to come back to that trip around the West that Beth and I took before leaving Montana. On part of the trip, we passed through the giant salt flats of Utah, where the white, salt-encrusted land stretches so flat for so far that people go there to try to set land speed records in amazingly souped-up cars that look more like fighter jets. Those salt flats exist because they used to be the site of ancient lakes. The lakes dried up, and as the fresh water evaporated, it left behind layers of salt, built up from the originally minute amounts that were slowly concentrated as the water disappeared.

When we irrigate in dry places, a very similar process leaves salt behind in our farm fields. It seems counterintuitive that dumping fresh water on soil would make it salty—after all, it's not salt water that we're using for irrigation. But even fresh water, including rain, surface, and groundwater, contains some dissolved salts. When you water a plant at home, those salts get added to the pot, and when water soaks through to the bottom pan and evaporates, the salts get deposited on the tray you've placed to catch the overflow. Over time, you will start to see the residue of those salts building up; it usually looks whitish on the terra-cotta pot. The same happens when salty seawater dries on your skin after you spend a day at the ocean.

The problem for farmers in dry regions is that the water they add via irrigation doesn't typically flush the system out from the bottom. Instead, thirsty plants suck the water up and release it through their pores, or water evaporates out of the soil without even going through the plants. In both cases, the salts dissolved in the water get left behind. They don't evaporate, and accumulate in the soil. The same residue that forms on an overflow pan beneath your houseplant begins to make the soil salty. And salty soils are a real problem for plants.

In high-enough concentrations, salt can kill plants outright. If you water your houseplants with seawater, they will die. But even in relatively low concentrations, salt interferes with a plant's ability to take up water. The problem arises when the soil is salty but the inside of the plant is not. Under these conditions, there is an osmotic gradient that pulls water from the places where water is fresh (inside the plant) to places where it is salty (in the soil). Osmosis is not an intuitive process, but you are probably familiar with it anyway. Drop a raisin in water, the salty interior of the raisin will draw water in, and the raisin will puff up. Take a long bath, and the salts in your body will pull relatively fresh bathwater in through your skin, making your fingers prune. Water flows from where it is relatively fresh to where it is relatively salty.

Why is this a problem for plants? Remember that plants get water by transpiring water out of their leaves. As water moves from the leaf to the air, it creates a pressure gradient which pulls more water out of the soil, into the roots, and up through the plant. I described it earlier as sucking up liquid through a straw. This process has to combat gravity, which is pulling the water back down. In salty soils, there is an additional force against the upward flow of water—osmosis. Water flows from fresh to salty environments, and if the soil is salty, it exerts pressure in the opposite direction the plants need it to go, pulling water out of the roots into the soil. So it's a tug of war—pressure gradients pulling water up as leaves lose water, and gravity and osmotic pressure pulling water down. If the osmotic force is strong enough, water can't be pulled up to the leaves. The cells that hold those tiny pores open will collapse, no more CO_2 will be able to get into the plant, and it will starve. Note that I did not say die of thirst. Plants need the water in the soil to eat CO_2, to turn it into sugars, starches, and other biological molecules. No water, no CO_2. No CO_2, no photosynthesis. No photosynthesis, no plants. No plants, no food for people.

Soil salinization is a real problem, and it will inevitably increase as irrigation of dry regions continues to expand. It's not

new. Scientists believe several Mesopotamian civilizations col-
lapsed because of soil salinization. But as the world's farms spread
into more and drier regions, fueled by the unsustainable extrac-
tion of groundwater, concerns about soil salinization are growing.
We are now capable of dumping much more water onto farms
than we were before we learned to tap groundwater, and we have
more people to feed. The amount of farmland threatened by this
process is hard to determine, but more than 20 percent of irri-
gated soils are increasing in soil salinity. The only way to get rid
of this salt is to flush more water through the soil out the bottom,
rather than leaving the salt behind while water evaporates back
into the air. But of course, we only irrigate in dry places, where
there isn't a lot of water to begin with. Using even more water to
flush out the salts will only lead to more rapid depletion of our
already stressed aquifers. The link between water and plants, and
thus between water and food, plays out in myriad ways.

My area of research, biogeochemistry, has its roots (pun
really intended this time) in farming. To grow crops, farmers
must understand the biology of those crops and also the temperature, light, rainfall, and soil in which those particular crops
grow best. Hence "bio" meets "geo" meets "chemistry." If you've
read this far, it will come as no surprise that biogeochemistry
underlies the greatest environmental challenges humans face.
That certainly applies to the biggest of all—climate change. For
the last part of this chapter, I want to spend a little time on the
way that climate change is likely to affect our access to water and
thus how it will affect our access to food.

———

Earlier in this book, I spent a chapter working through how we
use computer models to test and codify our ideas about the

climate system. I argued that our models are pretty good at identifying broad-scale patterns, and thus we are very certain of how warm the world will get given a certain amount of CO_2 put into the air by fossil fuel combustion and deforestation. But our models aren't nearly as good when it comes to understanding the future of rainfall. In general, climatologists are confident in the conclusion that most dry places will get drier, and most wet places will get wetter. But different models come to different answers about how much wetter or drier, which is a good sign that we have not reached consensus on what the future of rainfall has in store.

Nonetheless, the future of rainfall is obviously important, and knowing that drier places will get drier while wetter ones will get wetter is cause for concern. These changes will dictate where we can grow crops, whether the Amazon dries up into a grassland, and whether forest fires continue to rage out of control across the ever-drier forests of the western United States. But for all the uncertainty in how precipitation will change as the world warms, there is one thing about which we are certain—the warmer it gets, the more water will evaporate from the land. In places that are already dry, this warming will mean a dramatically drier future to which we will have to adapt.

Over the oceans, more evaporation in warmer climates will result in stronger and/or more frequent hurricanes, storms fueled by the evaporation and rapid rise of water vapor from the ocean surface. But on land, more evaporation means something very different. It means less water available for use. To understand this point, let's go back to the Colorado River, which you'll recall doesn't reach the ocean because it is trapped behind giant dams and siphoned off to crops and people in five states even before it reaches Mexico. I've already mentioned that every drop of that water is accounted for, litigated, and subject to endless argument. Shortly before I wrote this chapter, farmers in Mexico were leading protests calling for the release of more water. Many people—scholars, security experts, and those in the military—see climate-change-driven conflict over water as a major threat in the coming decades.

With that in mind, let's take a closer look at the future of the Colorado River. It will surprise no one that if there is less rain and snow in the high upper reaches of the watershed, the river will have less water in it. But as I said, the future of precipitation in the region is not very well constrained. We do know, however, that the region has been getting warmer, and that absent a dramatic shift in fossil fuel emissions, it will continue to get warmer quickly. Since 1900 the region of the Colorado River watershed has already warmed by almost 2°F. The eastern flank of the Rockies will almost certainly warm by at least another 2°F, and probably more, by the end of this century. Recent work has shown that for each degree it warms, the upper Colorado loses about 10 percent of its water. Why? Because more water simply evaporates off the land surface before it ever reaches the river. And that means a lot less water to fight over, which seems certain to make the fighting more intense.

The issue of warming and water far transcends the arcane water laws that govern the western United States. Over a billion and a half people, more than 20 percent of the world's population, rely on water derived from glacial meltwater. Glaciers sit high in the mountains and feed mighty rivers, draining the Himalayas, the Andes, the Alps, and even the Canadian Rockies. The mountains hosting those glaciers have been described as the world's "water towers" [22], storing vast quantities of fresh water and providing a steady stream to lowlands even if rainfall is seasonal and increasingly unpredictable. But, of course, those glaciers are melting quickly in the era of climate change, even as the populations depending on them are growing. What will this mean for the availability of fresh water to those billions in the future? No one really knows the answer in detail, but "there will be less of it" seems like a very well-supported conclusion. And less fresh water where it is already dry is not a good thing.

I realize that the more one dives into these problems, the more overwhelming they feel. Water, carbon, nitrogen, and phosphorus. If we really are changing the global flows of all these things, do we have anything but misery to look forward to? Don't the lessons of the past tell us of bad things to come? I have struggled with this question for my entire career, and on my bad days, I admit I am not up to the task of staying positive. But there is a ray of hope—we actually know a lot about how to maintain the benefits of our innovations without having to live through the worst consequences. And so, for the last part of this book, I want to turn toward that positive side. I want to explore how we, as individuals, institutions, and societies, can use what we know about Life's Formula to manage the transition we are inevitably going through as world-changing organisms.

Before we get there, I think it's very important to acknowledge that while the global changes I've been discussing have negative consequences, they are also the foundations upon which modern society has been built. It's impossible to imagine today's society without increased access to the elements in Life's Formula. Because of our ingenuity, the human population has grown from two billion to almost eight billion, and the percentage living in chronic hunger and poverty has plummeted. Life expectancy has more than doubled over the same period—the average human could only expect to live some thirty years in 1920.

Why?

Fertilizer and irrigation have vastly increased our ability to produce food. Carbon-based fossil energy has provided the platform for scientific and medical advances. You can't run a lab without access to energy, chemicals, and equipment. No energy, no mass production and distribution of medicine and medical devices. Then there are computers, which now use almost one percent of the energy that fossil fuels provide. Without computers we wouldn't have climate models to guide our airplanes safely through the air, molecular medicine, cell phones, and myriad other pillars of our global economy.

To understand the point, the next time the power goes out—because of a storm, a mistake by a utility, or just a squirrel that stepped on the wrong wire—think about how utterly dependent we are on a system built on the control of these elements. We use carbon-based energy to keep the nitrogen- and phosphorus-rich foods in our fridge cold, our house warm, and our transportation humming right along. The only thing that would really force you out of your house is a lack of water—which is treated and pumped to your house using the same carbon-based energy as everything else, and then treated as sewage using the same sources of energy. For those who have access to these elements, the basics of life are a given, only temporarily absent during some crisis. For those without, this elemental scarcity is what underlies the depths of poverty and creates the first barrier to emerging from it.

At the same time, the story of our world-changing predecessors and the science done to explore the unintended consequences of the very incomplete list of human successes noted above illustrate that something must change. Because we play by the same rules as all other organisms, business as usual will precipitate major changes to Earth's climate system and to a multitude of ecosystems upon which we have come to rely. As we warm coastal waters and pollute them with nitrogen and phosphorus, corals will die and disappear, potentially taking with them a quarter of the fish species on Earth. The Great Barrier Reef in Australia, the world's largest reef, is dying from overheating. That's bad for the fish, and it's also bad for the people who eat them. Fish provide almost 20 percent of the protein humans eat, with much lower impacts on biodiversity than other types of food production. On land, warmer and drier air feeds catastrophic forest fires that choke people and everything else. Already poor air quality kills millions of people a year, and near-permanent fire seasons will likely ramp that number up. Open any newspaper, and the unintended consequences of our elemental innovations jump from the headlines.

In my opinion, one of the great failures of the conversation about this dichotomy between the benefits and drawbacks of our modern way of life is that it is presented as an either/or situation. Either we go backward and give up the benefits of fossil fuels, fertilizer, and water infrastructure, or we go forward into global catastrophe. My own thinking was very much in this vein when I started learning about these issues. I longed for a world unperturbed by human footprints. In some ways, in my heart, I still do. But I know in my head that's not a reality. There will soon be ten billion people on Earth. They will need energy, food, and water. And by getting it they will inevitably need to manage the elements in Life's Formula. The time when humans were few enough to have only a minimal imprint on the planet is gone. But the future does not have to be one of unintended consequences.

PART III

A Way Forward?

8

Biogeochemical Luck

ONE OF THE MOST troubling comments from my students is that my class "Environmental Science in a Changing World" is "too depressing." I get it. It is depressing. I've struggled to see a way forward. I am an environmental scientist—someone who uses the scientific method to observe and understand the changes humans are making to the environment around us and to understand the consequences of those changes. But I am also an environmentalist, someone who values the natural world and seeks to minimize the adverse effects of human activities. Like many environmentalists, I have spent a lot of time looking backward. What I mean by this is that I often dream of returning to an imagined, idyllic past, when humans had a light footprint and nature was left to be as it was. A lot of scholarship has shown that this idealized past is a fantasy—at least for the last several thousand years, humans have had a big influence on the world we live in, and places that European colonists deemed empty hosted vast civilizations with profound ecological impacts [23]. My research and that of others in many fields have shown just how dramatic our alterations are. Despite this, the fantasy is a powerful one: a world left to its own devices, for us to enjoy, marvel in, and leave in peace.

But there are almost eight billion of us on the planet. We rely on altering the flow of the elements in Life's Formula and enjoy the benefits these elements bring. What's more, we will need more energy, more nutrients, and more water to build a more equitable future. So I've slowly abandoned my dream of returning to the past. We cannot go back in time when humans were not the managers of the Earth system. At first, I had nothing to replace this idea with, and it was then that my classes were the most depressing. But the more I've thought about the dictates of Life's Formula, the more I've come to think there is a way forward. Not to a future where we don't manage the Earth system, but to a future where we manage it more wisely. Nowhere is the need for wiser management more pressing than for carbon and climate change. I'll focus on this first and come back to the future of life's other elements in the next chapter.

So—carbon. Yes, we've changed the carbon cycle. Yes, the world is warming. Yes, the chemistry of life and of our planet will make those changes inevitable as long as we continue to create elevated levels of greenhouse gases in the air. Instead of focusing on the problem, though, I want to home in on the (potentially) life-saving difference between us and our predecessors.

Recall that for the cyanobacteria, whose proliferation meant the rapid rise of oxygen-generating photosynthetic organisms, there was no way to grow without pumping oxygen into the environment. The efficient photosynthetic reaction taking place in their cells has to produce oxygen. The land plants were similar constrained—their proliferation made planetary changes inevitable. The carbon stored in their bodies and their relentless attack on rocks to liberate nutrients to fuel growth eventually pulled enough CO_2 out of the air to precipitate massive changes in climate. Of course, humans play by similar physiological rules. We eat plants and animals that are full of carbon-rich compounds, break down those compounds to release energy, and breathe out CO_2. There is no way around that—it is baked into our chemistry. Fortunately for us, and in stark contrast to our world-changing predecessors, it's not the unalterable chemical reactions inside our bodies that are causing the global carbon cycle to change. The impact of our internal chemistry on the global carbon cycle is limited because there aren't nearly as many humans as there are of our world-changing predecessors. Plants make up more than 80 percent of all the living matter on Earth. Bacteria make up 12 percent. Humans? About one-hundredth of one percent. Our internal metabolism isn't a big deal—there aren't enough of us to matter.

In contrast to the other world-changers, we use vast quantities of carbon-based energy outside our bodies. This energy is released by burning fossil fuels to heat our homes, run our cars, and manufacture our goods. It is the use of this external energy that is the major driving force in climate change. That difference—between our internal and external use of energy—opens the door to a different future.

Let's compare those two types of energy use—the internal "burning" of food calories and the external "burning" of fossil energy to power the rest of our lives. The almost eight billion humans on this planet need a lot of energy to fuel our bodies. Each of us needs about two thousand kilocalories of energy a day to stay alive. Eight, and soon to be ten, billion people, each needing two thousand kilocalories a day, is a lot of energy required for our internal metabolism. But that number is dwarfed by the energy each of us uses to move our cars, heat our homes, keep the lights on, and make all the stuff we consume. That's more like fifty thousand kilocalories a day per person—a ratio of twenty-five to one. For the average American, who uses way more energy on things like cars and big houses than the average person in the world, that ratio is almost one hundred to one.

This back-of-the-envelope comparison makes it clear that our global domination of the carbon cycle isn't really about the number of people (though of course that matters). It's about the energy those people use and the emissions of CO_2, methane, and other greenhouse gases that come with producing that energy. That is why North Americans, who account for only 5 percent of the human population, emit 18 percent of all greenhouse gas. In contrast, Africans, who account for 16 percent of the global population, are responsible for only 4 percent of total emissions. We all need the same number of food calories. It's the rest of the things we do that really make the difference in terms of how much we impact the carbon cycle.

This difference from our world-changing predecessors—our need for societal, rather than biological, metabolism—is actually a saving grace. It means there is a path forward. It is a path whereby we get the energy we want to run our society in a way that does not emit greenhouse gases to the atmosphere and heat the planet at such a dizzying pace. It's a path that was not open to our predecessors. For every additional cyanobacterium or tree, there is a heightened impact on the carbon cycle. That doesn't have to be true for us.

———

How do we move down a different path? It's a question I think a lot of people ask, on a personal and societal level. We know in some way that the benefits afforded to many of us, particularly the relatively wealthy in relatively wealthy countries, are sending the whole planet careening toward a very difficult climatic future. The average American uses more electricity in a week than the average Kenyan uses in a year, and emissions of greenhouse gases differ similarly. We know that it is inequitable that the poorest people, who have done the least to cause the climate problem and have the fewest resources to adapt to it, are going to suffer the most for the emission-heavy lifestyles of the wealthy. And we know that even the wealthy will face the consequences of climate change, as will all other creatures on Earth. The news is replete with catastrophic stories—wildfires, floods, and hurricanes—all in some way linked to the human-caused climate changes that have really just begun. Some people worry more, and others less, but almost everyone sees change is happening. Coupled with that observation is panic, malaise, or both. And an overwhelming feeling that nothing can be done.

But a lot can be done. We know almost everything we need to in order to produce the energy we need without the greenhouse gas emissions we don't want. A lot has been written on this topic across an almost-endless list of disciplines. The bottom line is that we need to eliminate fossil fuel combustion, which accounts for around 70 percent of human-caused greenhouse gas emissions. While some emissions come from deforestation, agriculture, and cement production; fossil fuels are the biggest cause. Technological solutions to providing energy without carbon emissions, such as wind, solar, and hydroelectric (as well as nuclear) sources, have been endlessly discussed and promoted or dismissed by pundits, politicians, and academics alike. So too have the paths to implementing these technical solutions and their societal, economic, and political barriers. Regardless of your politics or personal opinions, the changes needed to transition away from business as usual can feel so overwhelming, so beyond individual control, that it is tempting to just ignore the problem and deal with the more immediate demands of daily life.

I want to take a different tack and imagine what daily life might look like without fossil energy. It's not a return to the era of the horse and buggy. In fact, I think you'll be surprised at how similar it looks to today and how close we are to getting there, although it will certainly take a concerted effort. I'm going to start with the center of daily life—the home. Since I'm the one writing this book, I'll start with my own home. My rather average American house is as good a place as any to understand the opportunities and challenges we all face in the transition to an energy-rich but low-emission future.

The house, shared with my wife Beth and daughter Phoebe, was built in the 1920s. It is about 2,500 square feet: a thirty-by-thirty-foot foundation topped by two stories and an attic. A chimney rises up the south wall next to the driveway we share with our neighbors. Their house looks just like ours, as does our neighbor's house to the north, and the house after hers. I think the house was built in the 1920s equivalent of suburban sprawl. Like most people's houses, ours is the center of our lives and thus the nexus for our use of life's essential elements. I want to explain how the house worked when we moved in and how it works now, as an example of how we can decouple our use of energy from our use of fossil-fuel carbon.

When we moved in (in 2007), the house probably functioned a lot like yours does today. Electricity was supplied via endless miles of transmission lines and transformers that run along almost all the streets in the world. Most of us rarely notice or give any thought at all to these wires until they are knocked over in a storm. If you start to look, though, you will find yourself amazed by their ubiquity and will marvel at these thin metal threads that literally stitch modern society together [24]. The electricity that flows to our particular house is produced by power plants far from us. The exact locations vary over time, but they are all within ISO New England, the name for our "regional transmissions organization" tasked with providing electricity to consumers. The emissions from those power plants don't come out of our chimney, but we use the electricity and thus are responsible for the carbon emitted during its generation. Where we live, about 50 percent of the electricity comes from burning natural gas, about 30 percent comes from nuclear power, and the rest comes from hydropower and, increasingly, wind and solar. Burning gas creates a lot of greenhouse gas emissions, both from the combustion and from all the methane that leaks out between the well from which the gas is extracted and the place it is burned. Remember that methane is a potent greenhouse gas, and that "natural gas" is mostly methane. If you count the leaks, burning natural gas causes almost as much climate change as burning coal.

Nevertheless, since the Environmental Protection Agency doesn't really count the methane leaks, according to official statistics, the Rhode Island electricity grid produces lower emissions per unit of electricity than most of the other regional grids in the United States. Despite our "relatively clean" grid, in 2007 our family's consumption of electricity accounted for about a third of our emissions. The biggest electricity users were the fridge, followed by the dryer, the dishwasher, other appliances, and computers (we didn't have air conditioners, which also use a lot of electricity). The Rhode Island electricity grid, like many grids in the country (and the world), is getting cleaner as the proportion of renewable, non-greenhouse-gas-emitting energy sources (such as wind and solar) feeding into it grows. Thus, emissions from our electricity consumption are falling a little each year. To accelerate that fall, my family could either use less electricity or rely more heavily on renewables (or both). We chose a path that actually increased our electricity consumption. But before explaining why that's a good idea, I have to take you through the rest of the house.

The next third of our family emissions in 2007 came from heating the house during the winter. About once a month between late October and early May, a tanker truck would pull up in front of the house, and the driver would run a long hose to an intake pipe in our driveway. He would pump oil into a two-hundred-fifty-gallon tank that sat, greasy and rusted, in the corner of the basement. Filling it would take only a few minutes. The tank fed an old furnace that burned oil to boil water, which then flowed as steam through asbestos-insulated pipes across the open basement ceiling before turning sharply up into the exterior walls. Burning this oil produced CO_2, which went up the chimney into the global atmospheric commons, where it will reside for longer than our house will be around. Of course, we had no interest in releasing CO_2; indeed we had no interest in carbon at all (well, I did, and Beth and Phoebe have to hear about it endlessly, but we had no interest in *releasing* CO_2). Our goal was a well-heated house that was comfortable. In truth, the house wasn't that comfortable. The steam radiators were either on, in which case it was too hot, or off, in which case it was too cold. Much of the time it somehow managed to be both. For those of you living in cold climates with steam radiators, I'm sure you know what I mean. Nonetheless, we were a lot more comfortable than we would have been without the furnace.

The furnace had a rating on it that said it was 85 percent efficient, and each year when someone came to service it, they said it was "about that." Eighty-five percent sounds pretty good, but none of the technicians were able to tell me what it actually meant. After digging around online, I figured out that the rating meant that 85 percent of the energy contained in the chemical bonds of the oil—bonds between carbon atoms originally formed by photosynthetic organisms—was converted to heat we could use. The other 15 percent was lost to "the environment," which in our case meant the basement (which was therefore nice and toasty) and up the chimney. Because the steam pipes from the radiator went through the exterior walls of the house, lots of heat (maybe as much as a third of the total energy contained in the oil) was lost to the outside through the minimally insulated walls. And because the pipes were in the exterior walls, the radiators were against those walls, too—usually under windows. That meant about another 5 percent of the heat meant to warm the rooms was leaving the house behind the radiators, heating the outdoors. In case you're not keeping track, that means only about half the energy stored in the oil was making it into the rooms it was intended to heat. But what can you do? We need heat, right? Well, yes. But there are much more efficient ways to get it. I'll come back to that after rounding out our home greenhouse gas budget.

The last third of our family's greenhouse gas emissions came not from our house but from our leaving it—mostly in the cars sitting in our driveway. At the time we had a relatively new Toyota Prius hybrid, which got almost fifty miles per gallon, and an old Subaru station wagon that got half that. Luckily for me, we were within walking distance of my work, but Beth is a doctor and had to be at the hospital at all sorts of odd hours, so she usually drove. Beyond work, we both spent a lot of time in the early years driving to and from Phoebe's school, especially before she was old enough to take the bus. All this added up to about fifteen thousand miles a year driven by the two cars combined. That's less than the average household in the United States (thirteen thousand miles per adult), but Rhode Island is very small, so we didn't have very much distance to cover in our daily routine. Nonetheless, driving these fifteen thousand miles per year produced about as much greenhouse gas as heating our house all winter, or using electricity all year.

I want to reiterate that our house in 2007 was probably very similar to hundreds of thousands of homes in the United States. Bigger than some, smaller than others, but basically a poorly insulated box kept warm in the winter by fossil fuel combustion, with appliances and lights powered by fossil fuel–generated electricity. And leaving the house—at least most of the time—required gasoline to move our two rather generic cars.

When we moved into the house, we started to ask whether we could get what we wanted—a warm house in winter and mobility—without getting what we didn't want, namely greenhouse gas emissions. While we couldn't eliminate our emissions (at least not without renovations outside our budget), we could cut them dramatically. And in so doing, we could prepare ourselves for a day when they could drop to zero. The first step came with the house itself. Built in 1920, its thin walls were insulated patchily with blown-in cellulose (basically shredded newspaper). The windows had been replaced by the previous owner, but they hadn't done much else to insulate, so each poorly insulated windowpane was surrounded by four-inch uninsulated gaps. We learned all this because we did three things. We asked for an energy audit (which most states subsidize), we hired someone to take infrared pictures of the house, and we did a blower door test. The pictures come from a camera that sees infrared radiation rather than visible light and thus can see hot and cold patches on the walls. These helped us find places where we needed to put in more insulation. The blower door test was much less high-tech. A guy came to the house, put a giant fan in the front door, sealed up the rest of the door with plastic, and closed the windows. We went around with a little smoking pencil and watched the smoke dive into nooks and crannies in the walls, which told us there were leaks that needed to be sealed.

This information allowed us to come up with a plan with a contractor. The house desperately needed a paint job—but instead of just painting the cracking shingles, we pulled them off, put some insulation on the exterior walls, and installed new siding. The attic was a mess. It had been half renovated by the previous owners so people could sleep up there, but the ceiling was low and the space uninviting. We knocked out the drop ceiling so that we could insulate the roof with a lot of spray foam, and we also rebuilt the ceiling higher and thus improved the value of the house (and the livability). Like most old houses in Providence, the basement leaked a bit, so we put insulation down there that could get a little wet and kept the insulation slightly away from the walls even though it would have been more energy efficient to sprayfoam that, too. We did an energy analysis that suggested that new windows wouldn't give us much energy bang for our buck, so we left our somewhat leaky, but not super old, windows in place and insulated around them.

All these renovations helped the house keep in heat and made it more comfortable. But far more importantly, they allowed us to change our heating system and get rid of that old oil tank. With the house better insulated, not super well, but around what code is for a new-build house in 2022, we could stop using that same combustion reaction that consumes organic material (oil) and combines it with oxygen to release energy and CO_2. Instead, we could use electricity to simply move heat from outside to inside the house.

To understand how we did this, you need to understand another key invention you may have never considered—the heat pump. I guarantee you have one in your house, and that you depend on it continually, because your fridge is a heat pump. If we are going to slow climate change, heat pumps also have to be the future of home heating.

How do they work? At the back or at the bottom of your fridge, there are two coils. If you're like me, those are the things you forget to brush even though the manual says you should. Anyway, inside those coils is a refrigerant that is very efficient at absorbing heat. Part of the coil goes into your fridge or freezer. The refrigerant goes into that coil and evaporates as it makes contact with the air in the fridge. This absorbs heat from the air in the fridge and cools that air down. The refrigerant is then circulated outside the fridge, where it is compressed back into a liquid. The compression requires electricity, and when the gas is compressed, it releases heat into your kitchen, away from the food in the fridge. Thus, the fridge works by moving heat from a place you want to cool (the inside of the fridge) and dumping it where you don't mind it (the kitchen).

Another heat pump you're familiar with is an air conditioner. It works in exactly the same way as your fridge does, except now your house is the box you are trying to cool. The refrigerant in the coils of the air conditioner absorb heat from inside the house, move it outside, and dump it there. Electricity is required because you're moving the heat from the inside (which is cooler) to the outside (which is warmer). You're sort of pushing uphill. But you're not creating any heat; you're just using an electric pump (and the chemical properties of the refrigerant) to move heat from one place to another.

How does this relate to home heating? Well, heat pumps can work in both directions. An air conditioner moves heat from where it's cold (inside the house) and dumps that heat where it's already warm (outside the house). By moving some of the heat from the inside to the outside, the air conditioner makes your house a little colder and the outside a little warmer. But what if you turn the air conditioner around on a cold day? That's what heat pump heaters do. The refrigerant circulates outside in the cold air, but the chemical properties are such that even the heat contained in the cold air is enough to make it evaporate. Then that refrigerant is pumped into the house and compressed, which releases the heat into the house. Heat pump heaters extract a little heat from the cold outside air, making the outside even colder, and move that heat into the house, making the house warmer.

The reason I labor through these details is that moving heat from one place to another is much more efficient than creating heat by burning fossil fuels. Using heat pumps means you can keep a house warm using much less energy, and therefore creating fewer greenhouse gas emissions, than if you were burning oil or gas in the basement. Remember, we don't want the side effects of greenhouse gas emissions. We only want a warm house. We can use what we know to get the latter without the former. Even more importantly, the energy for heat pumps can be supplied by electricity. This, in turn, can be generated without emissions as well.

Today our house has no oil tank or furnace in the basement. Gone are the asbestos-lined pipes running up to each room, and gone are the big clunky radiators that took up so much floor space. Instead, in a narrow sliver of backyard, there are six air conditioner compressors. Each one feeds a small copper tube that brings refrigerant to a head in each room. Inside the head, that refrigerant releases heat and then circulates back to the outside to pick up more heat. The system is quiet, without knocking and thumping the way steam radiators do, and works incredibly well. One day when it was ~15°F outside (as cold as it's been since we moved in), the air coming out of the head in the house was 140°F. The units are more efficient when it's warmer, and it's almost always much warmer than 15°F. But even on those coldest days, they are as or more efficient than our old oil furnace. And they are superefficient in the fall and spring, those long-shoulder seasons when it's cold enough to want some heat but not cold enough to really need a lot. All in all, the conversion to heat pumps reduced the greenhouse gas emissions released from heating our house by roughly half. Why not by 100 percent? Because those heat pumps use electricity, and in Rhode Island that electricity is coming in large part from natural gas consumption.

There isn't a lot we can do about that right now, but options are opening up, which I'll discuss in a minute. But I want to drive home the main point—the key to lower emissions from your house (or from buildings in general) is the electrification of home heating. Lose the furnace and stay warm. Get what you want (a warm, comfy house) without what you don't (a rapidly warming planet). Exploit the fact that, unlike our world-changing predecessors, most of our energy demand is external and not inexorably linked to the polluting byproducts of our cellular function.

One obvious question about all this is, How much does it cost to electrify? That's clearly an important question, since if only millionaires can afford to insulate their houses enough to use heat pumps, it will be impossible to reduce our collective emissions. Let me start by saying that for new construction, buildings with heat pumps cost the same or less than buildings with furnaces. And if you're going to own the building for more than a few years, the lifetime cost of ownership means any new construction that uses only electricity is almost certainly the cheaper option. But you have to start with a design that has no onsite combustion as a goal. Luckily, more and more architects and contractors realize this is the future, so it's getting easier to find people who know how to do this. If you're building a new house, please ask!

For renovations of old houses, it's more complicated because it very much depends on the house you have. Most of the time, from a purely financial standpoint, deep energy retrofits like ours make sense only if you can couple the energy reductions with other improvements (for example, we wanted to renovate the attic anyway, so it was an opportunity to super-insulate the roof). Big renovations simply won't pay for themselves in a few years the way an LED light bulb will. For example, we needed to paint the house, and adding insulation roughly doubled that cost, but the siding we installed was prepainted and was supposed to last twice as long as a regular paint job. So rather than pay $15,000 for a paint job that would have lasted ten or fifteen years, we paid $30,000 for one that would hopefully last twenty. The attic insulation cost another $12,000 after rebates, but we wanted to renovate the attic anyway, so adding insulation was just an opportunity that made that space way more comfortable. Finally, the heat pumps cost about $23,000, substantially more than the $5,000 or so a new furnace would have cost to install. Rhode Island offers zero-interest loans for energy-efficient heat pumps, loans which we paid back over five years. Plus, the house now has air conditioning in every room, whereas it had none before. That increases the value of the house for resale more than what we spent. Finally, and this is no small point, the house is far more comfortable than it was before in both winter and summer. Given how much time we spend at home, that's a big deal.

It's important to recognize that, for a lot of people, spending $45,000 on energy retrofits is out of the realm of possibility.

That's also true for renters since they can't make decisions about major infrastructure changes. There are companies striving to make changes cheaper, using high-precision laser scanning of houses to build external insulation molds that can be snapped onto the exterior in couple of days rather than a couple of months. Those are gaining traction in Europe, and there have been some pilot projects in the United States as well. These are great steps, but for now the transition to a fully electric home-heating sector is expensive for many. It will depend on those who can afford it going first in order to lower the cost for everyone else. It will also depend on government incentives to help lower the cost of entry and eventually regulation to limit the sale of gas furnaces.

While these sorts of steps are out of reach for many, it is also true that for a lot of people, the people whose homes are the biggest and who therefore have the highest emissions, these renovations are not cost prohibitive. Many people spend that kind of money renovating bathrooms and kitchens. It's a matter of priorities, but it's not a matter of possibility. The richest 10 percent of people on this planet are responsible for over half of all greenhouse gas emissions. Surely they can afford to go first, and by so doing, can drive the costs down for the billions of people who want well-heated homes and will benefit most from averting catastrophic climate change. It can be done.

Electrification is needed beyond the home, of course. For many people, this means giving up on the nineteenth-century internal combustion engine that powers their cars and switching to an all-electric vehicle. Right now, as with home renovations, electric cars cost more than their internal combustion competitors, but prices are falling rapidly, and a recent analysis found that, if you consider all the savings on fuel and repairs, electric cars already cost less than gas cars over the lifetime of the car [25]. As a result, most car companies are planning for a fully electrified future. Some governments are planning to ban the sale of new combustion-based cars; the United Kingdom plans to do so by 2030, and California, by 2035. My bet or hope is that by then there won't be a need. Electric vehicles will be so much better than their gas-burning counterparts that consumers will simply want them more. As I write this in 2022, electric cars already accelerate faster and drive better than gas cars. Their range is nearly five hundred miles per charge. They don't recharge in the time it takes to fill a tank of gas, but charging times are dropping quickly. In another decade, and likely less, it won't even be close.

Just like running a home, electrifying vehicles vastly increases efficiency. Recharging an electric vehicle still emits greenhouse gases because producing electricity isn't emission-free, but the emissions from charging our electric car are equivalent of emissions from a car getting over one hundred miles per gallon (the average car on the road today in the United States gets just under twenty-five miles per gallon). Just as with home heating, as the electricity grid gets cleaner or as people sign up for more emission-free electricity, emissions from charging electric vehicles will fall further. Remember, we want the ability to move around without the emissions. We're on the way there.

I want to come back now to the electricity that we do use—since after installing all those heat pumps in our house and buying an electric car, we have increased our electricity consumption, not diminished it. Because heat pumps are so efficient, our total greenhouse gas emissions have gone down. The same is true for our electric car. But we still use electricity from the grid, and, as I mentioned, that electricity comes almost entirely from burning natural gas. What can we do about that?

Of course, there are ways of producing electricity without producing greenhouse gas emissions. Hydroelectricity and windmills are the oldest of these technologies, and windmills are now spreading rapidly across the world, both onshore and offshore (where the winds are stronger and people complain less about the views). More recently, forms of solar electric generation, like the photovoltaics that are popping up on roof-tops, parking lots, and various open areas, have dropped so much in price that in 2020 the International Energy Agency declared these panels the cheapest form of energy generation on Earth.

There are also solar thermal installations, which work well in sunny regions. These are giant arrays of hundreds or thousands of mirrors, which concentrate sunlight on a small area and use the concentrated heat to melt salt. That molten salt can then be used to boil water and drive a steam electricity turbine in the same way that burning coal is used to produce steam and spin a turbine (without the greenhouse gas emissions). This has the advantage of working at night, after the sun has gone down, since the salt can be kept hot for hours after it is melted. Other ways of producing greenhouse gas–free electricity include nuclear fission (though that produces radioactive waste that no one has a great plan for disposing of and also presents a weapons proliferation threat) and nuclear fusion, a process that slams high-energy hydrogen atoms together to produce helium (the same process that fuels the sun). The latter has been the holy grail of electricity generation for some time, as it produces no radioactive waste and relies on the most abundant element in the universe (hydrogen). The problem is that no one has been able to figure out how to make it work without using more energy to power the fusion reaction than the useful energy it produces. However, a group of physicists recently estimated that fusion might be technologically feasible within a decade. That's been said before, but if that's true, it could certainly be a game-changer. Regardless, the major point here is that we have options for making electricity without greenhouse gas emissions, but that won't get us where we need to go unless we electrify everything—and move away from fossil fuel combustion for things like heating and industrial processes. I'm going to stay focused on homes and buildings, because that's something individuals can work on themselves.

A homeowner has relatively little control over what electricity is available to them, but many places allow you to sign up to purchase 100 percent renewable electricity. It often comes at a small premium, but even that becomes parity or a slight savings as renewable electricity generation gets cheaper. The idea is that if a whole bunch of people agree to pay that premium, the electricity provider can then make a deal with a large-scale solar or wind installation to buy that power instead of buying it from a natural gas or coal-fired power plant. Because it's cheaper to build large-scale installations than to put solar panels up on houses, that's a more affordable way to go for many people, and it's the only way to go for renters or people whose roofs don't face the right way to maximize sunlight.

Of course, renewables like wind and solar photovoltaics have one big problem—they are intermittent. Even if you electrify your home heating system or dryer, those emission-free sources of electricity aren't powering your house all the time. In the case of solar power, panels make less electricity when it's cloudy and none at night. Windmills don't spin when the wind isn't blowing. While it's technically possible to buy batteries for your house, charge them during the day, and use them to run a home overnight, it's not yet affordable for the vast majority of us (though battery prices, like solar panel prices, are falling fast).

Besides, batteries don't store all that much energy (yet). Thus, the best most of us can do right now is electrify our houses so that there is no onsite combustion, make our houses as efficient as possible, and subscribe to renewable electricity. As I mentioned earlier, even doing that is likely to drop your house's emissions by about 50 percent. Buy an electric car, and you'll shave 30 percent or more off the remainder. Couple that with replacing all light bulbs with LEDs (which use about 10 percent or less of the energy of incandescent bulbs) and replacing appliances with energy-efficient models when they are ready for replacement, and you're well on the path to zero emissions. Life won't change very much, except that your house will be more comfortable and your car will be more fun to drive.

Importantly, I don't want to suggest that individual action alone will be sufficient to slow climate change. The roles of governments, industry, and the private sector will all be critical. We can't slow, and eventually stop, climate change without major changes in the large-scale societal incentives to produce emission-free energy. Indeed, many people argue that an emphasis on individual action distracts from large-scale political and societal changes [26]. But—particularly for homes and for vehicles—there are things individuals can do to prepare for a cleaner power supply, and these things will reduce emissions directly while government creeps at its petty pace. Yes—we need regulatory action that moves us away from fossil fuel power generation. We will also need technological breakthroughs. Deforestation must end, and agriculture must be reformed since both are relatively big sources of emissions. Cement production must also change since cement is made from limestone ($CaCO_3$) and making cement removes a CO_2 from that $CaCO_3$ and releases it to the atmosphere. I'm not trying to downplay the complexity—there is plenty of it. I'm simply saying that, for many of us, particularly the wealthiest 10 percent of us who produce over half the world's emissions, there is a lot that we can do right now, and the rest is not impossible. It all comes back to that big difference between us and our world-changing predecessors. We can get the energy we need without the byproducts we don't want. Time, and not very much time, will tell whether we have the will to capitalize on that critical advantage.

There is no doubt that climate change, driven by our need for energy and the resulting change in carbon circulation around the planet, is the biggest environmental challenge ever faced by humans. But we do have an out—an out provided by the fact that the link between carbon and much of our energy use can be broken.

I hold on to this idea, because most days climate change can make me feel very hopeless. Sometimes it seems that, if we're going to slow this catastrophe, we're going to have to go back to living without all of the benefits of modern society. Turn off the lights, give up travel, shiver in the winter, and roast in the summer. That's just not true, and at the most fundamental level it's not true because, unlike our world-changing predecessors, our reliance on carbon as an energy source for our way of life is not the same as theirs. Cyanobacteria and plants altered the carbon cycle because of their internal metabolism. We did because of all the other things we use fossil fuels for—our external metabolism. And there isn't a need for the energy for our external metabolism to be carbon based and emit greenhouse gases.

Humans have built so much, changed the planet so much, and done it so fast. If you took all the stuff we produce—the buildings, roads, dams, and more—and put it on one side of a scale, it would outweigh the mass of every living organism on Earth combined. We emit ten times more mass of CO_2 to the air each year than we produce cement and steel combined. The possibility that we could continue to thrive while eliminating this impact on the carbon cycle would be truly remarkable. In my view, this achievement, should it come to pass in time to avert catastrophe, should be celebrated as one of the greatest human successes of all time. Yes, there are a lot of financial, societal, and technological barriers to overcome. But at least we are not constrained by our fundamental metabolism the way our world-changing predecessors were.

Having sounded that hopeful note, I want to turn to the rest of the elements in Life's Formula—nitrogen, phosphorus, and, of course, the elements in water. The constraints they impose are not so easily avoided, precisely because our need for them is related to our internal metabolism and the internal metabolism of the things we eat. We use the vast majority of H, O, N, and P for growing food. This presents a very different challenge from the one of energy and carbon. There is no way to grow plants (or the animals they feed) without these elements. As a result, there is no way to feed us all without affecting their global and regional flows through the environment. The constraints of biology are quite a bit more challenging than convincing people to take climate change seriously. For nitrogen, phosphorus, and water, the best we can do is improve our practices and minimize waste. But, in my opinion, that should be enough to avoid the worst consequences of our inevitable reliance on these essential elements. Thus, this too will be a hopeful chapter, albeit one with more unknowns left to resolve.

9

Some Remaining Puzzles

THE FIRST FEW DAYS I worked in a lowland tropical forest, at the La Selva Biological Station in Costa Rica, were some of the most amazing, but uncomfortable, of my life. I know it's a cliche to say, "it's not the heat, it's the humidity," but it really was the humidity. No break at night, no break when it's raining, no break at all. Everything dripped. The towering trees were draped with plants growing from their branches, and these epiphytes were themselves covered in mosses and lichens. Most of the leaves in these wet tropical forests have evolved little funnel tips that help shed water, so the leaves drip nearly constantly, even when it isn't exactly raining. People think of tropical rainforests as hot, but they aren't much hotter than my hometown on a summer day. They are just much, much more humid.

I don't want to be overly dramatic—I wasn't roughing it— and the verdant forest was a wonder to behold. The field station was well equipped, and there was good food, clean water, internet access, and lots of comforts many people don't have. I just wasn't used to the climate. Neophyte that I was, I brought my leather hiking boots and a leather belt for my fieldwork. Both began to rot after a few days. So did the leather case for the fancy compass Beth had loaned me, the one she got as a geology prize in college. It didn't survive the trip. If it hadn't been for the computer lab, which was air-conditioned for the sake of the computers, I imagined I would shortly begin to rot, too.

It turns out that all this rotting—what we dress up scientifically as "decomposition"—offers a window into how rainforests work. I want to use it as an entry point for the conversation about how human environments differ from rainforests, a conversation we explored a bit when thinking about phosphorus, and how we can use that knowledge to better manage the elements in Life's Formula.

There are a lot of reasons scientists like working in tropical forests. They are interesting because they take a lot of CO_2 from the atmosphere, they are host to about half the species that live on land, and they provide food, fuel, and fiber to millions upon millions of people. They are also just really inspiring places to be. Walking through the forest while listening to the birds and monkeys, even if they are maddeningly hard to spot, has been a life-changing experience for me. Yes—much of my research time is spent getting rained on while digging holes in the ground to collect soil samples. But the profusion of life around me as I muck around in the dirt is like nowhere else on Earth. The more you look, the more you see. A single acre of land in a tropical forest typically has more species of trees than are found in all of North America.

But if you're not a botanist—and I'll be honest, I definitely am not—it can just look like an endless wall of green. I've never taken a good photo in rainforest, and I've taken thousands of photos. Professional photographers do better, but even they struggle unless they are in a clearing, near a river, or in an airplane. There is a reason every nature calendar has a shot of a coral reef and a gorgeous mountain range but no rainforest pictures except ones of birds, jaguars, or monkeys, usually by a river. A day in the forest means a day without seeing the sky—it's blocked by layer upon layer of leaves. Some forests have eight square feet of leaves for every one square foot of ground. And those leaves have one purpose—to capture as much sunlight as possible to fuel photosynthesis. The forest is so good at capturing light that only about one percent of the light that hits the top of the forest canopy makes it down to the ground. It is always dark down on the ground, even when it's not raining.

At first glance, it is a bit puzzling how the forest manages to build so many leaves. As we've explored, many tropical forests sit atop really lousy, old, infertile soils. Some of those soils have been sitting in place for so long that most of the nutrients derived from rock have long been leached away. And for phosphorus, the most important of the rock-derived nutrients, the iron-rich remaining soil is like a magnet that holds it tight. Often when we try to measure the phosphorus available to plants, the data come back as a long column of "n.d."s, which mean "not detected." Sample after sample with no available phosphorus we can detect, but a verdant forest that seems to have no problem getting the phosphorus it needs.

It's a different story when humans cut the forests down and try to farm. Without fertilizer, people can produce only a few years of crops before the fields are essentially useless. How do the trees manage to build and maintain so many leaves, year after year, while people fail so quickly when they try to do the same? Tree leaves and crops both need nutrients: nitrogen, phosphorus, and all the other elements that allow plants to capture CO_2 and grow. If eons upon eons of high rainfall have flushed out almost all of the nutrients that were supplied by the underlying rocks, how do these forests manage to grow so much on so very little, whereas farmers can barely grow anything?

One reason that I gave earlier is that roots and fungi efficiently recycle the nutrients that fall to the ground in dead leaves. Here, I want to discuss a second reason, the chemistry of the leaves themselves. Science is replete with fancy instruments, mass spectrometers than can count individual atoms, and cameras that measure hundreds of colors (not just red-green-blue), but measuring the chemistry of fallen leaves is decidedly low-tech. It starts with four plastic posts and some window screening hung between them to catch the fallen leaves before they hit the ground and really start to rot (Figure 16).

The chemistry of fallen leaves turns out to be very different from that of the green leaves up above, which are packed full of nutrients. The dead leaves are nutrient poor. Trees start dissolving leaves before the leaves die, and pull critical elements back into the wood before a leaf falls. This reduces the amount they need to gather from the soil—they can reuse the same atoms over and over. In tropical forests like the Amazon, where phosphorus is often the element in Life's Formula most in demand relative to supply, it is phosphorus that the trees pull back the most. But the phenomenon exists worldwide. The spectacular colors that mark the arrival of fall across the middle latitudes occur when plants dissolve the green, nitrogen-rich photosynthetic machinery in their leaves and pull atoms back into their stems for use again in the spring.

While Amazonian soils are typically poor in nutrients (like phosphorus) derived from rocks, they tend to have quite a bit of nitrogen, at least relative to other essential atoms. As a result, the energetic cost of pulling nitrogen back out of their leaves isn't really worth it, and many Amazonian trees are not particularly efficient at conserving nitrogen. Why conserve something that's abundant and relatively easy to get? It is the opposite in

FIGURE 16. A team of biogeochemists from the lab on their way out to set up collectors to catch fallen leaves. Science in the rainforest is always happier when it's sunny! Photo courtesy of Brooke Osborne.

temperate forests. There, nitrogen tends to be scarcer than phosphorus (again, relative to the tree's needs). Perhaps not surprisingly, if you analyze fallen leaves in New England, you'll find that oaks, maples, and birches are most efficient at pulling a lot of nitrogen out of their leaves before letting them go—in a wild profusion of color—in the fall.

Environmentalists often talk about nature in contrast to people—and marvel at the efficiency of recycling in the natural world—nothing goes to waste. But I don't think we're all that different from plants. They can be wasteful, too—they just don't waste things that are in short supply. Plants are conservative with the things that are scarce, but less so with the things that are more abundant. This example of tropical trees being more conservative with phosphorus and temperate trees being more conservative with nitrogen is just one of many.

There is another, even more remarkable, example of the ways plants conserve life's two most abundant atoms: H and O. As you might expect, trees in a rainforest aren't very efficient at using water. To state the patently obvious: it rains a lot in a rainforest. But some rainforests have long dry seasons, and there the trees are more efficient in their water use than trees in rainforests that get rain twelve months a year. In fact, there are some rainforests that have such long dry seasons that the trees lose their leaves and regrow them when the rains come. Bone-dry, dusty places with no leaves are probably not what you think of when you think about the tropics, but travel to the Pacific coast of northern Costa Rica in March, and you'll see a landscape almost as leafless as the one you see in New England in January. This adaptation helps the trees conserve water and nutrients when it's too dry for the leaves to do much photosynthesis.

Of course, rainforests, even those with long dry seasons, are not the places where H and O are the scarcest. For extremes in what we call water use efficiency, you have to go to the desert. In the desert, cacti and other succulent plants are so water efficient that they have mostly abandoned photosynthetic leaves. Leaves lose too much water, and evolution has turned the leaves of cacti into thorns that serve as defenses rather than as photosynthetic machinery. The thorns are thus brown, packed with rigid compounds to make them sharp, and lack the green, nutrient-rich enzymes that carry out photosynthesis. Photosynthetic material is instead found in the surface of the stems, which is why the whole cactus looks green (as opposed to trees, whose stems are brown and don't photosynthesize). In addition, many desert plants have evolved a special kind of photosynthesis in which they open their stomata at night to let in CO_2 and then close them during the day, when they use sunlight to fuel photosynthetic reactions. Tropical trees don't do this—they usually have plenty of water. Trees in the tropics generally invest in retaining phosphorus. Temperate trees don't do it either—they invest in retaining nitrogen. But in the desert, where water limits growth, plants invest in retaining water.

The key point I want to make is that when some elements are scarce, evolution selects traits to use that element efficiently. But ecosystems tend to be less efficient with things that are present in abundance. The key to success is conservation and recycling, but the push for both comes only with scarcity. With this in mind, let's turn to humans and think about the way we use these same elements. Once again, we have the chance to learn from our world-changing predecessors and to change our behavior before scarcity or unintended consequences force our hand.

———

Let me start with a short history of the twentieth century through the lens of biogeochemistry. Fossil fuels and their carbon-based power provided humans with what appeared to be limitless energy, which enabled vast movements of people at unimaginable speeds, wars of horrific and unprecedented destruction, and the ever more energy-intensive (and greenhouse gas–emitting) lifestyle of the modern world. As the population grew, the biggest wars ended, and life expectancies began to grow, the lack of the other elements of Life's Formula became increasingly apparent. The energy of fossil fuels was turned toward the problem of nitrogen fixation (through the Haber-Bosch Process), phosphorus mining, and irrigation. By the mid-twentieth century, humans had emerged as the first organism in the history of life on Earth to change the abundance and flow of all the elements in Life's Formula. Not just carbon and nitrogen (cyanobacteria) or water and phosphorus (plants), but the whole suite: HOCNP.

As a result of these innovations, as I write this in 2022 the human population has grown to almost eight billion, the CO_2 concentration of the atmosphere gone up by around 40 percent since the Industrial Revolution, and as a result the global temperature has increased by over 2°F. There is at least twice as much nitrogen and phosphorus available on the continents as there was in 1900, and there is more water in reservoirs than in rivers. We are reveling in our success. Scarcity, where it persists, stems from social and political, rather than biophysical, causes. This is the first time in the history of the world where this has been so universally the case for one group of organisms above all others.

Not surprisingly, in the face of this unique abundance, we have become uniquely inefficient. The nitrogen fertilizer we dump on farm fields is mostly lost before it reaches the crops it is intended for, and even more is lost before it reaches our mouths as food. Once it passes through us, it is not recovered. We don't lose quite as much phosphorus on its way in because phosphorus is slightly less mercurial than nitrogen, but we don't do very much at all to conserve it on its way out. And water? Well, we grow water-inefficient crops in the desert, spraying mist into the air so that it evaporates before it hits the ground.

In short, we're a long way from mimicking the trees and the cacti.

For much of the past several decades, environmentalists have been pointing out the downsides of this linear, and highly inefficient, economy. They point to ecosystems and talk about how little is wasted and how much is recycled in them. That's true, in a way, but I think humans have been acting pretty reasonably given the circumstances. Abundant, cheap energy has made obtaining water and nutrients easy—so we haven't invested very much in conserving them. We're not really all that different from other organisms. They, like us, invest more in retaining those things that are hard to obtain. People use fossil fuels to make nitrogen fertilizer through the Haber-Bosch Process. We use fossil fuels to mine phosphorus from Morocco, China, and Florida and ship it where it's needed. We use fossil fuels to pump water from the Ogallala to grow corn in Kansas and build canals to siphon off water from the Colorado River to grow lettuce in Arizona. Our energy is subsidized, cheap, and reliable. As a result, so too are the nutrients and water they help us obtain. It's no surprise we haven't been very conservative.

The problem, of course, is that the environmental consequences of our inefficiencies are changing our climate, our waterways, and our ability to live well in the future. So what can we do to become more conservative? I'll start with a caveat: there is no perfect solution. In the last chapter, I suggested that, with energy and carbon, we pretty much know how to get to a better solution. There are certainly technical, social, and political challenges to be overcome. But there is little doubt that we can eventually overcome them. It's just a race against time since we have very little of it left to avert a truly catastrophic climate future. But for the other elements, we can only see a part of the path to lessening the unwanted consequences of our activities. So, for the rest of this chapter, I will focus on them and the challenge of becoming conservative (in the ecological sense) in the twenty-first century.

Of all the ways humans use H, O, N, and P, food production and consumption are the biggest. Given this, it makes sense to finish the book by talking about the other great sustainability challenge of our century: how to feed eight billion people now—and ten billion by mid century—in a way that produces nutritious food and minimizes human changes to the flows of the elements in Life's Formula.

Once again, I'm going to start with my house as a focal point to explore the differences between the way we do things now and the way we could do things. But as with carbon and energy, the problem cannot be solved by end users alone. There have to be systematic reforms at every level of society. So, after thinking through the things individuals can do at the household level, I'm going to move back through the supply chain. I'll give you some examples from industrial-scale farms, where most food is produced, and talk about the ways in which scientists and farmers are working together to better manage the atoms that flow through them. From there, we'll go to some of the poorest cities in the world and a remarkable organization working to bend the linear trajectory of nutrient loss back into the circular economy of a tropical rainforest. More than half of all humans live in cities, and the number is growing every year. If we want a system that reuses waste in the way a forest captures nutrients from leaves before they fall, we need to understand how to make our cities part of a cycle rather than the conduit through which hard-won atoms escape our grasp.

Since I'm starting small and building outward, I'm going to start in the fridge, specifically my fridge. Ours looks a lot like everyone else's. Its biggest difference from the average American's is that we gave up eating meat a few years ago—although I still sneak an occasional spicy Szechuan chicken from a restaurant near my parents' place in New York. But meat, particularly red meat, is where I want to start the path toward lowering human impacts on H, O, N, and P.

A few years ago, a paper came out in which the authors tried to estimate the total amount of matter stored in various kinds of life on Earth—plants, bacteria, mammals, fish, and so forth.

It was one of those rabbit holes scientists sometimes crawl into. Has anyone ever counted this up? How could one do it? What will the answer be? It wasn't a revelatory new method or the discovery of some far-off world. But it was a cool idea, nonetheless. I read the paper because I was intrigued by the question, and one thing really jumped out at me.

The authors found that if you placed all the people in the world on one side of a scale and all the livestock (such as cows, pigs, or chickens) on the other, the livestock would outweigh people by about two to one [27]. Humans eat so much meat that we have populated the planet with livestock that outnumber (and outweigh) us. The comparison is even starker in terms of absolute numbers—there are almost three times as many chickens on the planet (over twenty billion) than there are people, plus another billion cows and another billion pigs and goats (combined).

What does this have to do with H, O, N, and P? Feeding all those animals requires a lot of plants, plants that use a lot of nitrogen, phosphorus, and water. In some parts of the world, animals are grazing on lands that otherwise wouldn't be arable. In that case, they gather nitrogen and phosphorus from the plants they eat, concentrate them in their bodies, and provide those essentials to people. As long as there aren't too many animals, the soils aren't too infertile, and there is enough rain, it is a pretty sustainable system. But in other places, livestock graze on soils that are so poor that even a little nutrient removal by the animals slows the rate of plant growth. This is the story of the Amazon, where Brazilian beef is raised, slaughtered, and shipped around the world. And in drier regions, particularly where human populations are booming, the number of animals simply overwhelms the ability of plants to regenerate, leaving an ever dustier, less productive, and more denuded landscape over time.

More and more, however, meat production requires more than grazing land. For example, private range and grazing lands make up almost 30 percent of the area in the contiguous United States. That's a lot. But the more than ninety million cows roaming these lands rarely graze on wild plants for their whole lives. Instead, they are eventually shipped to feedlots to be "finished." There, they are fed a diet rich in corn and other nitrogen-, phosphorus-, and water-intensive crops. The crops are grown on farmland that only produces animal feed. Add that farmland to the grazing land, and we see that about half of the continental United States is used for growing cows. Globally, a third of all the land we use to grow crops is devoted to growing crops to feed animals.

This number is staggering, but there are a lot of people in the world, so you might argue that it is land well spent. Maybe. But the heart of the issue with eating animals, particularly cows, is that they are very inefficient in turning plant calories into meat calories. The exact conversion factor varies with the type of cow and the type of feed, but a good rule of thumb is that it takes ten calories of plants to make one calorie of cow. If the cow is grazing on plants we can't eat, then at least we're getting something out of a landscape we can't get food from (though we might do something else with it). But if the cow is eating corn or other crops that we grow specifically for them, we are using way more nitrogen, phosphorus, and water than we need— roughly ten times as much as we would if we ate those crops ourselves. Put another way, the United States (a population of 320 million) could produce enough calories to feed another billion people if it stopped feeding crops to animals and instead grew crops that people ate directly.

For this reason, the path to feeding people with less of a footprint on life's essential elements requires fewer cattle (and pigs, which aren't much more efficient than cows). It isn't necessary to get rid of all meat. But feeding crops to people rather than to animals has to be part of reducing our impact.

To illustrate this point, I have the students in my introduction to environmental science class do a very simple calculation. I ask them to figure out how much land it takes to feed eight billion people. I tell them how many calories a productive farm can grow each year on an acre of land. For simplicity, I have them assume every farm on Earth is highly productive (this is far from the case). I tell them how many calories the average person needs to eat per day, again simplifying but assuming one number for every person, regardless of age, sex, or size. But I have them do the calculation two ways. In the first, I ask them to assume that everyone in the world, all eight billion people, eat a "US" diet. That includes an average of fifty-five pounds of red meat per year. And I remind them that it takes about ten calories of crops to make one calorie of cow. Then I ask them to do the exact same calculation—but this time they assume that we are all vegetarian and we eat those crops ourselves rather than feeding them to cows.

The point of the calculation is not scientific certainty. There are all sorts of oversimplistic assumptions that wouldn't pass peer review. Still, the results are often eye-opening for the students, as I hope they will be for you. To feed the world a US diet, specifically a diet with as much beef as people in the United States consume, would require us to double the amount of land that is currently used for farming. That sounds like a lot, and it is. Let me try to put that doubling in context.

Humans already use almost 15 percent of Earth's land surface to grow food. In total, that's an area about the size of South America. We use another 35 percent, about the size of Africa, as grazing lands. That's 50 percent of the ice-free land surface used to produce food. How on Earth (literally) could we double that? Clearing the rest of the Amazon wouldn't be enough, nor would it be enough to clear the last rainforests from Africa in addition to that. Almost all of the arable land in North America, Europe, Russia, and Asia is already in use. Climate change will someday render more land in the north farmable but will drive other land out of production, so on average warming won't open up much new land. There just isn't anywhere to go.

In contrast, feeding the world a vegetarian diet would require about half the land we currently farm. The reason for this is simple—it is inefficient to feed food to animals, and then animals to people. Eating the food ourselves would free an area the size of Brazil from farming, allowing it to return to some other uses. Again, the idea here isn't the specifics. The exact numbers aren't right. But the idea is. Eating less red meat, particularly cows and pigs, is probably the most important thing humans can do to reduce the unintended consequences of our food system. Every bushel of crops we feed to animals rather than people means we've spent all that effort to gather these elements and flushed them down the toilet.

Of course, lots of people prefer eating meat. It has great cultural importance in many (but not all) places around the world. Anyone who has traveled to France, Brazil, or Argentina, to name a few countries where I've worked, has seen the centrality of meat in diets. Beyond cultural preferences, meat is a key source of protein for many people who have very little food security, and for them animals often serve as a caloric savings account that can be drawn on in times of need. In general, though, the world's poorest are not the people eating fifty-five pounds of meat a year the way the average American does.

Indeed, in the world's poorest, most food-insecure regions, people average only about seven pounds per year.

As I write this chapter, several possibilities for replacing meat are coming to the fore. Perhaps the most intriguing one, and most relevant to the themes of this book, sprang from the mind of Patrick Brown, professor emeritus of biochemistry at Stanford University. Brown, a pediatrician turned basic scientist, created the first high-profile open-access scientific journal (you don't have to pay to read) and invented the DNA microarray (a staple tool in any modern biology department). But for the purposes of this story, he is most interesting for the company he founded in 2011: Impossible Foods.

By now you've probably seen ads for the Impossible Burger or even tried one. We now serve them in the cafeteria of my university. I recently saw a twenty-foot neon sign for it in the Atlanta airport. It is one of many meat-replacement burgers, but it is unique in an interesting way—a way that relates directly to the atoms at the center of this book and links all the way back to cyanobacteria and nitrogen fixation.

As I mentioned earlier, nitrogenase, the enzyme that converts N_2 in the air into reactive nitrogen, evolved a long time ago, long before there was oxygen in the air. Which is a good thing, because nitrogenase binds irreversibly to oxygen, and once it does, it can no longer pull nitrogen from the air. What, you might ask, does that have to do with a fake meat burger? Stay with me. The link is fascinating.

In the chapter on humans and nitrogen, I spent a lot of time on legumes and how they build little nodules on their roots that house nitrogen-fixing bacteria. I said those "houses" kept out oxygen, which is important because oxygen binds and inactivates the enzyme that carries out nitrogen fixation. But

I didn't say how the houses kept oxygen out and that's the link to the Impossible Burger.

Legumes keep oxygen out of their bacteria "houses" by infusing their root nodules with a molecule called leghemoglobin. Chemically, leghemoglobin is very similar to the hemoglobin in our blood, which binds oxygen and carries it throughout our bodies. Leghemoglobin also binds oxygen. But its function is simply to keep oxygen away from those precious nitrogen-fixing bacteria and their nitrogenase enzymes. Which is why, if you cut a legume nodule, it turns red, just like our blood.

After a lot of testing, Brown and his team deduced that the thing that makes a burger taste like a burger is actually the cow's hemoglobin. So, when they set out to make a vegetarian burger that tasted like meat, they turned to the legumes. They implanted the gene that codes for leghemoglobin into yeast and used that genetically modified yeast as a key ingredient. Since that yeast was packed with leghemoglobin, they got a veggie burger that bleeds. I recently had one, and although the texture isn't exactly the same, the flavor would have fooled me easily. Why go to all that trouble? Because plant-based diets are the fastest way to reduce human impacts on H, O, N, and P. If people won't stop eating Big Macs and Whoppers, maybe they will eat Big Macs and Whoppers made of a meat substitute that bleeds.

OK—so less or (even better) no red meat is the first key. But the truth is that farming, even regular old farming, requires a lot of nutrients and water, and a lot of H, O, N, and P are lost before they reach our mouths. I mentioned in the chapter on nitrogen that for one hundred units of nitrogen applied to a farm field as fertilizer, only about fourteen reach our mouths if we're vegetarian, and only about four if we eat meat. The numbers probably aren't quite as bad for phosphorus, but they aren't wildly different. This makes meat a clear loser, but looked at another way, switching away from meat only gets us from losing 96 percent of the nitrogen embedded in the food before we eat it to "only" losing 86 percent. That's better, but it's a long way from efficient by any definition.

Where can we find an example of an efficient system? We can go back to the land plants. They do everything they can to retain nutrients and water where they grow. They invest the most in retaining the things that are scarcest. Right now, H, O, N, and P aren't that scarce for many people, so we're wasting them.

It all comes down to keeping essential elements where we want them and not where we don't. There is lots of room for improvement here. Modern technology allows farmers to add fertilizer at the time it's most needed, to optimize the number of plants per acre, and to test soil to make sure there is enough, but not too much, nitrogen. Genetic engineering, while controversial to some, can be used to make plants more nutrient efficient and maybe someday will allow biological nitrogen fixation in crops (like corn) that aren't legumes. If genetic modification via test tubes makes you uncomfortable, conventional plant-breeding for better nutrient retention is also underway.

Another reason rainforests are so efficient at cycling nutrients is that there are always plants in the soil—there is never a time when there is just bare dirt. But drive across any agricultural area after harvest and before planting, and it is a vast landscape of exposed dirt. Worse still, because of labor and price constraints, as well as spring muds, farmers often put fertilizer on in the fall. Those nutrients sit for months leaching slowly (or quickly) off the field even before crops are there to take them up. Farmers, researchers, and others are learning how effective winter cover crops, plants that stay on the field in the winter and are removed for spring planting, are at keeping fertilizer on the farm and out of waterways. Others are experimenting with perennial grain crops—which grow for several years and keep soil in place. One—Kernza—recently showed up in a box of cereal in my local supermarket. These and many other techniques are coming online, leveraging modern technology to increase food production while minimizing nutrient losses. No one wants the unintended consequences. Lots of people—farmers, scientists, and government officials—know nutrient losses are a problem and are working to avoid them.

It would take another whole book to talk about the future of farming and how we can improve the food-production systems upon which we all depend. For now, I want to introduce one person who is making a difference by understanding the link between farmers, society, and Life's Formula. A couple of years ago, my friend and colleague Lisa Schulte-Moore invited me to give a talk at Iowa State University. It's always an honor when your colleagues ask you to do this, and I was also excited because Lisa invited me to go "into the field" with her class. With this being Iowa, a state that is 98 percent managed land, we weren't going into a forest. We were going to a corn and soybean farm, one very much like the vast majority of farms in Iowa. Corn and soy are used for myriad purposes, almost none of which are feeding people directly. Mostly they feed pigs, cows, or ethanol distilleries.

The harvest was over by the time I arrived, and most of Iowa looked like a muddy, rutted mess—with fewer live plants per acre than anywhere but the driest deserts. Still, the farm where we were going was a little different. There were the ubiquitous leftover cornstalks, cut neatly at a uniform height by the combine that harvested them. But interspersed were narrow strips of native prairie grasses towering above the clear, rutted, muddy remains of the corn harvest. These strips had been designed and planted by Lisa, a team of scientists, and a very enthusiastic farmer named Tim.

Sitting with Lisa at Tim's kitchen table, I listened to them explain the idea. Iowa has fantastic topsoil, around sixteen inches of the most fertile land for growing crops anywhere in the world. At least, it did. About half of it was lost over the past century, as native prairie grasses were displaced by crops, tractors began to break soils apart, and drainage systems were put under almost every square foot of land to help corn grow better. If you're not from a farming region, you probably haven't heard of tile drainage. But under Iowa's farms run thousands upon thousands of miles of underground gutters that shunt water directly off the farm fields and into drainage ditches. Tile drainage keeps the fields from getting too soggy, but the water carries soil, and a lot of nitrogen and phosphorus, with it when it goes.

As a result of plowing, keeping soil bare all winter, and tile drainage, only about half of Iowa's original amazing topsoil is still in Iowa. The rest has flowed down the Mississippi River, carried off by accident. Every farmer knows that once topsoil is gone, there is no quick way of putting it back. We've known this since the dust bowls of the 1930s, and even further back than that. There is even a government agency called the Natural Resources Conservation Service that helps farmers work to slow soil erosion. But erosion keeps happening. Farmers are squeezed economically, and on a year-to-year basis it's cheaper not to deal with erosion, even if over the long term it's the end of the line for farming in Iowa and much of the rest of the world's breadbaskets. It's the end of the line because when the soil is gone, so too are the nutrients and the soil's ability to hold water for the plants to use.

Against this backdrop, Lisa, her team, Tim, and many farmers like him are exploring how to get the benefits of industrial agriculture—lots of food produced on relatively little land at low cost—without the drawbacks (lost nutrients, no habitat for anything other than crops, intensive use of pesticides, and myriad other ills). What they've learned is really interesting. On an experimental farm, Lisa's team took 5–10 percent of the land, the least productive land, and planted diverse strips of prairie grasses perpendicular to the flow of water. Then they measured how much soil and nutrients the farms lost and compared it to fields with no prairie strips. Surprisingly, even a small amount of prairie grass decreased the amount of nitrogen and phosphorus lost by the farm by something like 40 percent. Soil erosion as a whole was reduced by 90 percent. And the strips did something even more important. They began to build soil back up. They weren't just stopping the withdrawals from the soil bank account—the prairie plants were making new deposits.

The team moved beyond the first experiments, partnering with innovative farmers like Tim and with other landholders big and small to expand across the Midwest. The strips are not a panacea. Farms will still need nitrogen and phosphorus to replace the losses that come with the harvest. But at least the strips can staunch the bleeding, slow the loss considerably, and buy time to figure out further improvements.

Luckily, prairie strips are but one idea in a whole movement of so-called regenerative agriculture, agriculture that uses plants to conserve nutrients and soil at the same time that the plants produce food. After all, plants are what build soil in the first place—this has been true as far back as the colonization of land by plants four hundred million years ago. It makes sense that we can work with plants to make food *and* to build soil rather than using plants solely for food and using chemicals in place of soil.

Some of these regenerative approaches will turn out to be flawed. Others, and I'm betting prairie strips will fall into this category, may revolutionize the way we farm. I'm far from the only one betting on Lisa—she won a MacArthur "Genius Grant" in 2021. Lisa is a powerful spokesperson who is now working with some of the world's largest food producers to expand on the lessons learned from early adopters. Farmers are lining up to participate—demand is higher than what her team can meet. Lisa's work, and that of others like her, can help us move toward solving the problems that come with our success in manipulating life's essential elements.

As encouraging as this and other work on agriculture is, in order to operate a truly efficient system we will still need to find a way to recycle the atoms that leave the farm. There are two other major loss pathways—through us, and through our animals. I've already talked about reducing the numbers of animals, so let's focus on us. Most of us are aware of the problem of untreated sewage in economically disadvantaged countries. For example, in São Paulo, Brazil, a city of twelve million, only a little more than half of the sewage is treated at all. In Nairobi, Kenya, a city of more than four million, far less than half is. This is first and foremost a direct health threat. But even in most wealthy countries, where effluent passes through a chemical or biological treatment that removes pathogens, nutrients are not recycled. Sewage sludge is full of nitrogen and phosphorus, but none of it is made available to put back on farms. In my hometown of Providence, Rhode Island, the treatment plant uses bacteria to turn the nitrogen in the sewage into nitrogen gas and return it to the atmosphere in order to keep it out of the local bay. That works, but after all the energy that went into fixing the nitrogen through the Haber-Bosch Process and spreading it on farm fields, it's a shame to lose it back to the air. After treatment, our sewage sludge is trucked to an incineration plant, where the soggy mess is burned (another enormous fossil fuel expenditure) before the residue is spread over the landfill. Figuring out how to safely recycle this waste, waste that is packed with life's essential elements, is one of the major urban design challenges of this century.

Again, there are really promising, innovative solutions that people are working on. So let me introduce you to another colleague of mine—Sasha Kramer. Sasha and I got our PhDs from the same department at the same time. Unlike many of us who were just trying to keep our heads above water while completing our dissertations, Sasha made it clear that the social and equity issues that motivated her were going to be front and center all the way. I remember reading about large protests at a major international conference one morning, and then coming to work where people were circulating a photo of a woman in a pink feather boa and giant pink shades in front of hundreds of armed riot police. Yup—that was Sasha.

Though her PhD project was focused on organic agriculture in Oregon, Sasha's passion was the work she had started in Haiti, the most economically disadvantaged country in the western hemisphere. As a PhD student working on biogeochemistry, she was learning about the power of Life's Formula. As an advocate working in Haiti, she saw both devastation and opportunity for improvement in its management. Haiti has many challenges; two among them are most relevant here. The first is that there is very low soil fertility because of the soil erosion that followed the deforestation of the relatively steep landscape. Because most farmers don't have a lot of money and fertilizer is expensive, Haiti cannot grow enough food on the island to feed its population and is forced to rely on expensive imports. A second major problem is the treatment, or lack thereof, of human waste. In Port-au-Prince, and indeed in pretty much all Haitian cities, untreated sewage pours into open ponds that pose a human health hazard, release large quantities of nitrous oxide (a strong greenhouse gas), and eventually overflow into the ocean. And that overflow, as you probably guessed, takes the nutrients that are so desperately needed on the farms in the hills and disperses them into the sea.

Even during her PhD program, Sasha spent a lot of time in Haiti building a nonprofit* to work on these two problems simultaneously. Often sleeping in the office of her nascent organization, she started to raise money to build composting toilets in urban neighborhoods. She did experiments to figure out how long and under what conditions the human waste collected in these toilets needed to compost before the pathogens they contained were gone and the nutrients they contained could be safely put back on nearby farms. At her doctoral defense, Sasha brought a mason jar of her human-waste-derived fertilizer and made everyone in the room smell it. It smelled like soil, of course, since soil comes from decomposed organic material. But it was still quite a sight to see a bunch of members of the National Academy of Sciences gingerly open, and then sniff, Sasha's compost. She didn't stop there. Rumor has it that on one of President Clinton's visits to Haiti, Sasha jumped on a scooter, tracked him down, and tried to make him smell her compost, too. I haven't verified this rumor, but knowing Sasha, I don't doubt it.

Sasha's way of making the world a little better through an understanding of Life's Formula is small-scale, grassroots, and effective. But despite its small scale, it recognizes a reality that goes far beyond Port-au-Prince. More than half of humans live

* The nonprofit is called SOIL (Sustainable Organic Integrated Livelihoods).

in cities. Soon that number will be closer to two-thirds, as megacities sprout at unprecedented rates and existing cities swell. This transition has profound implications for all aspects of society, but it is an opportunity for the conservation of the elements of this story. Concentrating all those people in a relatively small place means recapturing the critical elements that flow through them is just a tiny bit easier.

Since we won't run out of nitrogen as long as we have access to energy, and there is currently enough phosphorus around (though of varying price), it's perhaps not surprising that reclaiming these elements has not been at the top of most people's minds, even those who worry about environmental issues. But the issue is starting to attract more scientific attention. A recent study in Sweden, for example, showed that a large fraction of farm fertilizer needs could be met through the recycling of human waste but that it would increase costs substantially. Perhaps we are not willing to bear those costs today, but given that urban sewage infrastructure lasts a long time, we'd be well served to start thinking about making it better now.

If nitrogen and phosphorus recycling is still too expensive for most to envision, hydrogen and oxygen are a different story. Of course, both are already scarce in arid regions, which are getting drier because of climate change. Both are getting scarcer for billions of people who rely on shrinking mountain snow-packs that once were annually refreshing bank accounts for dry summer months. As in nature, scarcity drives innovation in conservation. Los Angeles, for example, has treatment plants that turn sewage water into potable water; so do two cities in Texas and the Namibian capital city of Windhoek, which has done this since the late 1960s. Israel now reuses more than 90 percent of its wastewater. To date, this cactus-like ability to reclaim H and O has depended on fossil fuels—a saving of H and O at the expense of energy, carbon, and greenhouse gas emissions. But there is no a priori reason this has to be the case. In fact, even oil-rich Saudi Arabia, which leads the world in desalinizing salt water to produce fresh water, is adding solar capacity to help fuel the process. At the end of the day, our ace in the hole is that we can get energy without carbon, and we can use that energy to get the other atoms we need. It's only a matter of time.

How much time? For climate change and the consequences of our carbon use, not very much. We are locked in to substantial warming and sea level rise as a result of what we've already done. Both will be monumental challenges to face. Like many things in our ever more unequal world, the worst consequences will be faced by people with the fewest resources to adapt. This is the greatest injustice of our innovation. The benefits accrue mostly to a wealthy few, while those with fewer resources suffer the most egregious consequences. Some of the worst will arise from the interaction between climate change and drought. We are wildly inefficient with water in rich countries, so there is room for rapid improvement. But in poorer, arid countries where people depend heavily on local food production, these effects will be devastating.

We probably have a bit more time to deal with the global concerns around nitrogen and phosphorus. We can use energy to get the former without reliance on fossil fuels. Non-fossil-based nitrogen fertilizer will come, and we won't ever run out of nitrogen as long as we have energy. We still need to reduce the loss of nitrogen on its path from farm to mouth, but again there is lots of room for improvement and innovation. It's not hard to see those improvements continuing. Phosphorus is available only in finite amounts, so it will need to be recycled. But we're not running out this decade, and probably not even this century. As long as we can anticipate the consequences enough to change our behavior (not a given), our access to phosphorus is a big, but solvable, problem.

There is reason for hope, because unlike our world-changing predecessors we can see what will come to pass if we do not craft a better path. We want energy, not carbon. We want water, but we need it mostly for our crops, and we can use energy to get it in innovative ways. Food production does not have to become 100 percent efficient overnight. There are things we can do, from planting prairie strips to recycling sewage, to help keep nutrients where we need them. Nitrogen is abundant. Phosphorus is a trickier puzzle, but even some recycling will stretch out the window of time before our global reserves are depleted. There is a path forward. But it requires that we, like all organisms before us, reckon with the constraints imposed by Life's Formula.

Our transformation will need to be fast, and I'm sure it won't be perfect. It already hasn't been. But we know what the problems are. We are world-changing organisms, and like our predecessors, our success depends on access to life's essential elements. The twentieth century saw us increase that access to an unprecedented degree at an unprecedented rate, which produced unprecedented strides in human well-being. We've already begun to see the unintended consequences of our extraordinary success. There is no reason, or at least no scientific reason, why the twenty-first century shouldn't see us use our understanding to refine our energy and food production, learning from the elemental thread that ties us to our world-changing predecessors.

ACKNOWLEDGMENTS

THIS BOOK was first conceived in 2013 during a training program for midcareer environmental scientists called the Aldo Leopold Fellowship Program. I benefited enormously from this training, both personally and professionally, and have become a better scientist, writer, and thinker because of it. I am especially grateful to the small group of fellows who still meet regularly: Trevor Branch, Jill Caviglia-Harris, Claudio Gratton, Kevin Krizek, Lisa Schulte-Moore, and Jennifer Tank (collectively, the Woodchucks!). Your successes are a great source of inspiration for me, and I am truly grateful for the feedback, encouragement, and friendship I've received from you all over the years.

Long after it was conceived, but also long before pen hit paper, Nancy Pick convinced me that I could write a book, and Cornelia Dean convinced me that I shouldn't unless I absolutely had to. I am grateful to both of them for helping me learn to write more fluidly, concisely, and (I hope) effectively. Alison Kalett and Hallie Schaeffer at Princeton University Press have been inordinately patient with my endless delays, and similarly helpful in making my prose somewhat less clunky.

I am also grateful to several key mentors without whom I would never have become the person I am today. Kevin Sweeny and Gordon Levin introduced me to scholarship and writing as they guided my senior honors thesis in history at Amherst College. Jack Cheney and Tekla Harms introduced me to the wonders of geology and the science of the natural world. Peter Vitousek took me on as a biology PhD student even though I hadn't taken a biology class since the ninth grade, and taught me how to be a scientist and mentor. I am forever grateful to him, and to Pamela Matson and Oliver Chadwick, for showing me a whole new world and teaching me how to explore it.

Both of my parents died this year, and I have been thinking even more than usual about the role they played in helping me become who I am today. My mother adored science written for a wider audience, avidly reading and sharing the works of Richard Feynman, Stephen Jay Gould, and Raymond Smullyan. I still have issues of *Natural History Magazine* from the 1970s and 1980s with notations of articles I might enjoy neatly circled on the cover in her instantly recognizable hand. It was a thrill to first publish some of these ideas in *Natural History Magazine* and think of her. Although she descended into dementia long before this book came to fruition, my father read an almost-finished draft just before he died. A psychoanalyst who claimed not to have a scientific mind, he offered insights and ideas that belied this claim. Most importantly, their love and support in this and in everything I did allowed me to follow my interests and passions and know that they had my back.

Finally, my wife Beth and my daughter Phoebe have patiently endured my struggles with this project for way too long. By the time she reached sixth grade, Phoebe had heard several hour-long seminars about these topics at her school, on a trip to Antarctica, and during various monologues at the dinner table. Her patience with me and all things sustainability understandably occasionally wears thin, but her innate curiosity usually wins out. I am impressed by, and grateful for, her interest in pretty much everything, including the work described here. I hope that we as a society can make the changes we need to in order to ensure a sustainable future for her, her generation, and their children. Finally Beth, the love of my life, has taught me so much about thinking, mentorship, and the world in general that it is impossible to imagine this book, or any other part of myself, without her. It is a true gift to share a life with someone who loves to think about the world together. Thank you.

WORKS CITED

1. Priestly, S. J. *Experiments and Observations on Different Kinds of Air and Other Natural Branches of Natural Philosophy Connected with the Subject.* (2nd Edition) London: J. Johnson, 1775.

2. Canfield, D. E. *Oxygen: A Four Billion Year History.* Princeton, NJ: Princeton University Press, 2013.

3. Vitousek, P. M., and Howarth, R. W. "Nitrogen Limitation on Land and in the Sea: How Can It Occur?" *Biogeochemistry* v13 p87–115, 1991.

4. Redfield, A. C. "The Biological Control of Chemical Factors in the Environment." *American Scientist* v46 p205–221. 1958.

5. Dukes, J. S. "Burning Buried Sunshine: Human Consumption of Ancient Solar Energy." *Climate Change* v61 p31–44. 2003.

6. Arrhenius, S. "On the Influence of Carbonic Acid in the Air upon the Temperature of the Ground." *Philosophical Magazine* 41, no. 251, p237–276. 1896.

7. Box, G. E. "Robustness in the Strategy of Scientific Model Building." In *Robustness in Statistics,* edited by R. L. Launer and G. N. Wilkinson, 201–36. Cambridge, MA: Academic Press, 1979.

8. Kennefick, D. *No Shadow of Doubt: The 1919 Eclipse That Confirmed Einstein's Theory of Relativity.* Princeton, NJ: Princeton University Press, 2019.

9. IPCC, Arias, P. A., Bellouin, N., et al. *Climate Change 2021: The Physical Science Basis. Contribution of Working Group I to the Sixth Assessment Report of the Intergovernmental Panel on Climate Change.* Cambridge: Cambridge University Press, 2021. https://www.ipcc.ch/report/ar6/wg1/.

10. Gould, S. J. *The Flamingo's Smile: Reflections in Natural History.* New York: W. W. Norton and Company, 2010.

11. Berman-Frank, I., Lundgren, P., and Falkowski, P. "Nitrogen Fixation and Photosynthetic Oxygen Evolution in Cyanobacteria." *Research in Microbiology* v154, p 157–64. 2003.

12. Smil, V. *Enriching the Earth: Fritz Haber, Carl Bosch, and the Transformation of World Food Production.* Cambridge, MA: MIT Press, 2001.

13. Hager, T. *The Alchemy of Air: A Jewish Genius, a Doomed Tycoon, and the Scientific Discovery That Fed the World But Fueled the Rise of Hitler*. New York: Crown, 2009.

14. Vitousek, P. M., J. Aber, et al. "Human Alteration of the Global Nitrogen Cycle: Causes and Consequences." *Issues in Ecology* vi, p2–16. 1997.

15. Smith, C., Hill, A. K., and Torrente-Murciano, L. "Current and Future Role of Haber-Bosch Ammonia in a Carbon-Free Energy Landscape." *Energy and Environmental Science* 13, no. 31, p331–344. 2020.

16. Matson, P. A. *Seeds of Sustainability*. Washington, DC: Island Press, 2012.

17. Firestone, M. K., and Davidson, E. A. "Microbiological Basis of NO and N_2O Production and Consumption in Soil." In *Exchange of Trace Gases between Terrestrial Ecosystems and the Atmosphere*, edited by M. O. Andreae and D. Schimel, 7–21. New York: John Wiley & Sons, 1989.

18. Vitousek, P. M. *Nutrient Cycling and Limitation: Hawai'i as a Model System*. Princeton Environmental Institute Series. Princeton, NJ: Princeton University Press, 2004.

19. Wulf, A. *The Invention of Nature: Alexander von Humboldt's New World*. New York: Knopf, 2015.

20. Walls, L. D. *The Passage to Cosmos: Alexander von Humboldt and the Shaping of America*. Chicago: University of Chicago Press, 2009.

21. Rockstrom, J., W. Steffan, et al. "A Safe Operating Space for Humanity." *Nature* v461 p 472–475. 2009.

22. Immerzeel, W. W., A.F. Lutz, et al. "Importance and Vulnerability of the World's Water Towers." *Nature* 577, p364–369. 2020.

23. Mann, C. C. *1491: New Revelations of the Americas before Columbus*. New York: Vintage, 2006.

24. Bakke, G. *The Grid: The Fraying Wires between Americans and Our Energy Future*. New York: Bloomsbury USA, 2016.

25. Miotti, M., Supran, G. J., Kim, E. J., and Trancik, J. E. "Personal Vehicles Evaluated against Climate Change Mitigation Targets." *Environmental Science and Technology* v50, p10795–10804. 2016.

26. Mann, M. *The New Climate War: The Fight to Take Back Our Planet*. New York: PublicAffairs, 2021.

27. Bar-On, Y. M., Phillips, R., and Milo, R. "The Biomass Distribution on Earth." *Proceedings of the National Academy of Sciences of the United States of America* v115, no. 25, p6506–6511. 2018.

28. Berra, Yogi. https://www.goodreads.com/author/quotes/79014.Yogi_Berra.

29. Coleridge, Samuel. "The Rime of the Ancient Mariner." In William Wordsworth and Samuel Taylor Coleridge. 1798. *Lyrical ballads: with a few other poems*. London: Printed for J. & A. Arch.

30. Behringer, Wolfgang. *Tambora and the Year Without Summer*. Translated by Pamela Selwyn. English Edition 2019. Cambridge: Polity Press.

INDEX

Adams, Addie, 103
adenosine diphosphate (ADP), 118
adenosine triphosphate (ATP), 118
ADP. *See* adenosine diphosphate
agriculture: in Amazon region, 126–27;
    biogeochemistry and, 155; in Central Valley, California, 147–48; corn
    production, rates of, 110; fertilizer
    use in, 111–12; food production and,
    111; Green Revolution, 111; irrigation
    systems, 152–53; nitrogen fertilizers,
    108–14; Ogallala Aquifer in U.S.,
    149–51; phosphorus fertilizers and,
    127–29; regenerative, 204–5; rivers
    as water source for, 147; salty soils
    and, 153–55
air conditioners, 174–76
air pollution, 160; in U.S., increases in,
    134
algae, in oceans, 22. *See also* dead
    zones
alternative stable states concept, 138
Amazon Rainforest, 22; deforestation
    of, 124–25; fungi in, 125; in lowland
    regions, 122; nutrient-poor soils in,
    188–89
Amazon region: agricultural production in, 126–27; dams in, 146
amine group, 101

amino acids, 101–2
ammonia (NH_3), 102, 108
anoxia: early development of Earth
    and, 9; planets and, 16. *See also*
    "dead zones"
Antarctica, 6; measurement of carbon
    dioxide in ice cores from, 75, 85
Anthropocene: 4
aquifers: GRACE satellites, 151–52;
    Ogallala Aquifer, 149–51; under
    Sahara Desert, 149; wells and,
    150
archaea, 22
Archaeopteris, 44, 47–48
Arrhenius, Svante, 84–92
astrobiology, 5
Atacama Desert, 6
atmosphere, of early Earth, 16; carbon
    dioxide removed from, 51; oxygenation of, 48–49; oxygen percentages
    in, 16
ATP. *See* adenosine triphosphate

bacteria: carbon and, 103–4, 108; land
    plants and, 106; nitrogen and, 106;
    nitrogen-fixing, 106; in oceans, 22;
    purple sulfur, 25. *See also* cyanobacteria; soil bacteria
Barry, Jim, 75

rivers, as water source: for agricultural
production, 147; damming of, 146
RNA, carbon in, 6
rock-derived elements, 39–40;
through dissolving of rocks, 53; in
Life's Formula, 40
root systems: in land plants, 44, 46–47;
recycling of nutrients by, 188

Sahara Desert, 23; aquifers under, 149
soil salinization, 153–55
Saskatchewan, 142
Saudi Arabia, 142, 209
Schindler, David, 135
Scott, Sheldon, 57–58, 63, 65
sedimentary rocks, 16
seeds, of land plants, 43, 47–48
La Selva Biological Station, 185–86
sewage sludge, recycling of, 205–6,
208–9
sewage treatment, in Haiti, 206–7
Sierra Nevada region, 6
simulations, in climate change
models, 84
slow carbon cycle, 59, 62, 69; humans
and, 65–66
Smil, Vaclav, 110
soil. See soils
SOIL. See Sustainable Organic Integrated Livelihoods
soil bacteria, nitrogen fertilizer trans-
formations by, 114
soils: in Amazon Rainforest, 188–89;
carbon storage in, 50; on Hawaiian
Islands, 121–22, 125–26; in Iowa,
202–6; iron oxides in, 123; from lava
flows, 120; nutrient-poor, 188–89;
salty, 153–55; in tropical rainforests,
21–22, 186, 188–89

solar energy: incident on Earth, 77;
electricity from, 169, 180–82
soybeans, 126–30, 202; GMO, 129;
non-GMO, 129; phosphorus fertilizers and, 127–30
stomata, 142–43
succulents, 144, 190
sugars: carbon in, 6; respiration
and, 63
sunlight: land plants and, 43; in photosynthesis, 20
super phosphates, 131
surface water, 145–46
Sustainable Organic Integrated Livelihoods (SOIL), 207
Sweden, 208

tectonic plates, 52, 61–62
temperatures, global: climate change
as influence on, 70–73; after Industrial Revolution, changes in, 98.
See also climate change
thermodynamic laws, in climate
change models, 95
tropical rainforests. See rainforests

United States (U.S.): air pollution in,
134; energy use in, 167; Environ-
mental Protection Agency, 169;
Natural Conservation Service, 203;
Ogallala Aquifer, 149–51; phospho-
rus ore deposits in, 132; water pollu-
tion in, 134
University of Montana, 140
Urey, Harold, 50
U.S. See United States

vegetarian diets, 198–200
Venus, 16

ABOUT THE AUTHOR

Stephen Porder is professor of ecology and evolutionary biology at Brown University, and a fellow in Brown's Institute for Environment and Society. He also serves as Brown's Associate Provost for Sustainability, the first such position at an American university. His research focuses on carbon and nutrient cycling in tropical forests and agriculture, the possibility of and barriers to forest restoration, and the transition to a fossil-fuel-free society. Porder was graduated from Amherst College with a BA in history, the University of Montana with an MS in geology, and Stanford University with a PhD in biology. In addition to authoring more than seventy peer-reviewed publications, he has published in the *New York Times, Folha de São Paulo*, the *Providence Journal*, and *Natural History Magazine*. He is the co-founder of the podcast and radioshow "Possibly," which explores sustainability science and many of the themes discussed in this book. Porder is also a Leopold Leadership Fellow, a group of over two hundred environmental scientists dedicated to communicating science to a wider audience and translating knowledge into action. He lives in Providence, Rhode Island, with his wife Beth and daughter Phoebe.